Calculating Theoretical Yield Is *Not* Impossible!

Calculating theoretical and percent yield is a fundamental skill for the laboratory. This book primarily targets Organic Chemistry Laboratory courses at the high school or college and university level, as a supplemental resource to help students master this skill. It begins with simple examples from everyday life, demonstrates the importance of balancing the equation, addresses the role of the mole in these computations, discusses different types of liquids, considers the role of significant figures, and culminates with the planning of syntheses. There are suggestions for further reading as well as practice problems and questions to ensure mastery.

Calculating Theoretical Yield Is *Not* Impossible!

A Primer for an Organic Chemistry Lab Course

Michael S. Leonard

CRC Press
Taylor & Francis Group
Boca Raton London New York

CRC Press is an imprint of the
Taylor & Francis Group, an **informa** business

Designed cover image: Shutterstock

First edition published 2025
by CRC Press
2385 NW Executive Center Drive, Suite 320, Boca Raton FL 33431

and by CRC Press
4 Park Square, Milton Park, Abingdon, Oxon, OX14 4RN

CRC Press is an imprint of Taylor & Francis Group, LLC

© 2025 Taylor & Francis Group, LLC

ISBN: 9781032959726 (hbk)
ISBN: 9781032959719 (pbk)
ISBN: 9781003587408 (ebk)

DOI: 10.1201/9781003587408

Typeset in Sabon
by Deanta Global Publishing Services, Chennai, India

Access the Support Material: www.routledge.com/9781032959719

To Jeni

Contents

Preface

Calculating theoretical (and percent) yield is a fundamental skill for the laboratory. It is among the earliest lessons we learn in the lab, and it is also among the most persistent skills that we use time and time again. Most students learn this skill quickly and retain the ability to perform the calculations from one year of their education to the next. However, in over 20 years of teaching laboratory courses, I've observed that some folks struggle with this concept and never quite master it. This is especially unfortunate because, as long as you are creating substances in the laboratory, the need to perform these calculations never disappears, so theoretical (and percent) yield can be among those dreaded topics that haunt you from one year to the next—unless, of course, you tackle it head-on right now and make sure that you truly understand both the idea behind these calculations and the "nuts and bolts" of performing them in various situations. Just think of all the points you'll save on lab reports, lab quizzes, and lab exams if you can calculate theoretical and percent yield easily! The time spent now to gain a command of these concepts will pay great dividends in the future.

This primer is meant to help you do just that. We'll begin with simple examples from everyday life to illustrate just what theoretical and percent yield are anyway. These simple examples will also help us appreciate the importance of balancing the equation prior to any calculations. Then, we'll address the mole: what it is and why it is so central to these calculations. We'll discuss different types of liquids—neat versus solutions—and how to manage the calculations involving them. The role of significant figures will be considered too. Finally, we'll flip the whole process on its head as we learn how to plan the amounts of reactants needed for a synthesis. Each chapter has a small number of practice problems with fully worked solutions, and this primer ends with a set of 20 questions (again, with detailed solutions) to make sure you've mastered the concepts for your lab course.

I once had a student several years into their college career bemoan the fact that "Calculating theoretical yield is impossible!" I assure you that it is not. In fact, it is a relatively straightforward process once you truly understand the basic tenets and have practiced them in a variety of scenarios. This primer will help you do that. You've got this! Let's get to work.

About the author

Michael S. Leonard earned his B.A. in Chemistry from Goucher College in 1998 under the direction of Professor David E. Horn. He then transitioned to the University of Pennsylvania for his doctoral studies in the laboratory of Professor Madeleine M. Joullié. After obtaining his Ph.D. in 2003, he joined the faculty of Washington & Jefferson College, where he is Professor of Chemistry.

https://orgowithleonard.com

What is theoretical yield anyway?

Theoretical yield is a simple concept that you already understand intuitively based on your everyday experiences. Imagine that you visit a big-box furniture store and purchase a kit needed to make a table. You would expect the kit to come with the tabletop and four legs. Once you assemble the pieces, you'll have one complete table (Figure 1.1).

In this case, the theoretical yield is one table. That's the maximum number of finished products that you can make with the starting materials provided. Can you make less than the theoretical yield? Sure! For instance, if you break one of the table legs during assembly, then you won't be able to make a fully functional table, and you'll make zero tables instead of one. *However, you can never make more than the theoretical yield.* Using the pieces that you have in this kit unaltered, you simply cannot end up with two, three, five, or ten tables. It is just not possible.

The theoretical yield is the maximum amount of product that you can make with a given amount of starting materials (called reactants in chemistry).

Once we know the theoretical yield, we can calculate a percent yield, which is a measure of how successfully we did our job. If you used the tabletop and four legs to build one table, then you made 100% of the tables it was possible to build in this scenario. If we express this in an equation, we see that:

$$Percent\,Yield = \frac{amount\,of\,product\,obtained}{theoretical\,yield} \times 100$$

In the specific situation, we've been describing:

$$Percent\,Yield = \frac{one\,table\,obtained}{one\,table\,that\,could\,be\,built} \times 100$$

$$Percent\,Yield = 1 \times 100 = 100\%$$

DOI: 10.1201/9781003587408-1

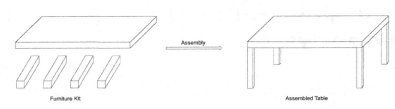

Figure 1.1 Table assembly from a kit.

If we built that one table successfully, then we did our job perfectly, and earned a "score" (a percent yield) of 100%. However, if we broke one of the table legs during assembly and therefore failed to make a fully functional table, our percent yield is 0%:

$$Percent\,Yield = \frac{zero\,tables\,obtained}{one\,table\,that\,could\,be\,built} \times 100 = 0\%$$

The percent yield is an assessment, on a scale of 0–100%, of how success-fully the product was created. We use percent yield to gauge our success because almost everyone has a very firm grasp of the fact that 100% is perfection and 0% is utter failure. As students, we are accustomed to earn-ing grades, so these percentages make a great deal of sense to us. You'd be thrilled to earn 100% on your chemistry exam, but you'd be mortified to earn 0%.

Now, let's consider a different scenario. Let's say you find a retailer where you can purchase tabletops and table legs separately. If you buy two tabletops and eight legs, then you have enough pieces to make two complete tables (Figure 1.2).

In this case, the theoretical yield is two tables. If all goes according to plan and you successfully build both tables, then your percent yield is 100%:

$$Percent\,Yield = \frac{two\,tables\,obtained}{two\,tables\,that\,could\,be\,built} \times 100 = 100\%$$

If you accidentally crack a tabletop during assembly, then you'll only be able to complete one full table, and your percent yield will be 50%:

$$Percent\,Yield = \frac{one\,table\,obtained}{two\,tables\,that\,could\,be\,built} \times 100 = 50\%$$

If you were really careless and broke both tabletops, then your project is a total failure, and you end up with zero tables, giving you a percent yield of 0%:

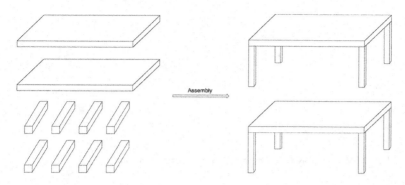

Figure 1.2 Assembly of multiple tables.

$$Percent\,Yield = \frac{zero\,tables\,obtained}{two\,tables\,that\,could\,be\,built} \times 100 = 0\%$$

The two scenarios above are very straightforward because the starting materials complemented one another perfectly. In other words, you had four table legs for every tabletop. Sometimes things don't work out quite so neatly. What if you placed an order for delivery but didn't receive everything you ordered? Imagine that the delivery truck drops off two tabletops and six table legs. Now your starting materials (or reactants) are not perfectly matched (Figure 1.3). You have enough pieces to build one complete table. Then, you'll have one tabletop and two legs left over. These aren't going to do you much good because they don't constitute enough pieces to construct another table. In this case, your theoretical yield is only one table. If you build it successfully, your percent yield will be 100%, but if you fail to assemble it, your percent yield will be 0%.

This example also introduces the notion of the limiting reactant (also sometimes called the limiting reagent). When the amounts of two (or more) starting materials are not perfectly matched, whichever one runs out first limits the amount of product that can be made. This starting material is therefore called the limiting reactant (or limiting reagent). In this case, you had enough tabletops to build two tables, but you only had enough table legs to build one table. The table legs run out first and, as a result, are the limiting reactant. It is the limiting reactant that dictates what the theoretical yield will be. In other words, the number of table legs told us that we'd only be able to make one table. We had an excess of tabletops, so the number of tabletops was really irrelevant to the calculation of theoretical yield.

The starting material that runs out first is the limiting reactant (or limiting reagent), and it determines what the theoretical yield will be.

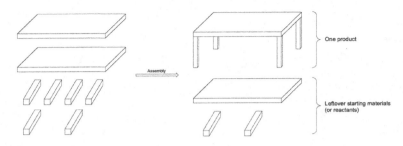

Figure 1.3 Table assembly from imperfectly matched components.

These examples may seem overly simplistic, but as we'll see throughout the subsequent chapters, the principles outlined above are exactly the same as those governing the calculation of theoretical and percent yield for chemical reactions. Since molecules are tiny, we'll usually be dealing with enormous numbers of molecules, even in small samples, but these ideas are scalable. It doesn't matter whether you are making 1 large product that you can see (such as a table) or 6.022×10^{23} minuscule products that you cannot see (such as molecules); the concept is the same either way.

CHAPTER 1 PRACTICE PROBLEMS

1. You go to a big-box furniture store and purchase a kit for a bookshelf. This kit comes with two long vertical pieces and three short horizontal pieces (Figure 1.4). These pieces can be assembled to form a bookshelf.

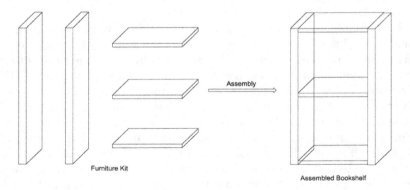

Figure 1.4 Bookshelf assembly from a kit.

(a) Given the assembly instructions above, what is the theoretical yield for this kit?

(b) If you successfully build one bookshelf, what is the percent yield?

(c) If you accidentally break some pieces during assembly and are therefore unable to complete a usable bookshelf, what is your percent yield?

2. Having successfully built a bookshelf that was very much to your liking, you return to the big-box furniture store to purchase the makings of additional shelves. In this shopping trip, you acquire the pieces shown in Figure 1.5.

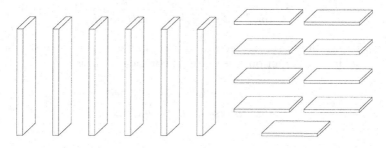

Figure 1.5 Selection of components for bookshelf assembly.

(a) Assuming that you plan to build bookshelves according to the schematic in Problem #1, what is the theoretical yield?

(b) What is the percent yield if you successfully build two bookshelves?

3. You tell a friend about the great deal you found on bookshelves, and they rush to the store to purchase the makings of their own DIY bookshelves. However, your friend has some confusion about the schematic. They purchase the parts shown in Figure 1.6.

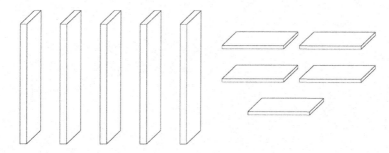

Figure 1.6 Selection of components for bookshelf assembly.

(a) If we assume that your friend will build bookshelves according to the original schematic given in Problem #1, which starting material (long vertical pieces or short horizontal pieces) runs out first?

(b) What is the theoretical yield of bookshelves?

(c) If your friend builds one bookshelf, what is the percent yield?

SOLUTIONS TO CHAPTER I PRACTICE PROBLEMS

1. (a) If you follow the assembly instructions you've been given, this kit has enough pieces to construct a single bookshelf. The theoretical yield for the kit is therefore one bookshelf.
   (b) If you do successfully build one bookshelf, then your percent yield is 100%:

   $$Percent\,Yield = \frac{one\,bookshelf\,obtained}{one\,bookshelf\,that\,could\,be\,built} \times 100 = 100\%$$

   (c) If you break some pieces and are unable to complete the bookshelf's construction, then your theoretical yield is 0%:

   $$Percent\,Yield = \frac{zero\,bookshelves\,obtained}{one\,bookshelf\,that\,could\,be\,built} \times 100 = 0\%$$

2. (a) You've purchased six long vertical pieces and nine short horizontal pieces. Since each bookshelf requires two long vertical pieces and three short horizontal pieces, you have enough parts to build three bookshelves (Figure 1.7). The theoretical yield is three bookshelves.
   (b) If you successfully build only two bookshelves, then the percent yield is about 66.7%.

   $$Percent\,Yield = \frac{two\,bookshelves\,obtained}{three\,bookshelves\,that\,could\,be\,built} \times 100 \approx 66.7\%$$

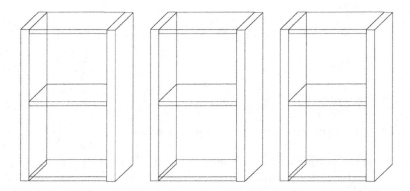

Figure 1.7 Assembled bookshelves.

3. (a) Your friend purchased five long vertical pieces. Since each bookshelf requires two long vertical pieces, your friend has enough long vertical pieces to make two bookshelves. However, we must also consider the other starting material: the short horizontal pieces. Your friend purchased five of these as well. Since each bookshelf needs three short horizontal pieces, there are only enough of these to build one bookshelf (Figure 1.8).

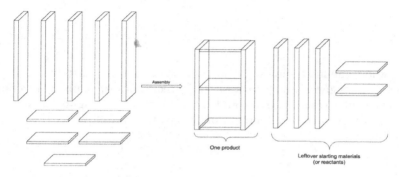

Figure 1.8 Bookshelf assembly from imperfectly matched components.

The short horizontal pieces run out first. These are therefore the limiting reactant.

(b) Though there are enough long vertical pieces to make two bookshelves, we only have enough short horizontal pieces to build a single bookshelf. The limiting reactant determines the theoretical yield because it runs out first. The theoretical yield is one bookshelf.

(c) If your friend builds one bookshelf, then the percent yield is 100%.

$$Percent\ Yield = \frac{one\ bookshelf\ obtained}{one\ bookshelf\ that\ could\ be\ built} \times 100 = 100\%$$

At first glance, this might seem counterintuitive. How can the theoretical yield be 100% if your friend didn't use all the pieces they purchased? To understand this, it is important to remember that theoretical yield is *not* based on using all of every reactant. It is based on using as much of the limiting reactant as possible. In this case, the limiting reactant was the short horizontal pieces, and there were enough of those to build only one single bookshelf. Your friend managed to do that, so they assembled the maximum number of bookshelves that they possibly could have. This is why the percent yield is 100%. Following the schematic and with the number of pieces available, no one could have made any more bookshelves than your friend did. They achieved the best possible outcome.

Chapter 2

A balanced chemical reaction
The key to everything!

A reaction equation shows the starting materials (or reactants) and the products for a given transformation. It is a recipe of sorts with the reactants on the left of a reaction arrow and the products on the right. To continue with our analogy from Chapter 1, our recipe was as follows:

$$\text{Tabletops + Table Legs} \rightarrow \text{Tables}$$

However, this reaction is unbalanced. In other words, it does not show the numerical relationship between tabletops, table legs, and completed tables. A balanced reaction equation shows those numerical relationships, as in the example below:

$$\text{1 Tabletop + 4 Table Legs} \rightarrow \text{1 Table}$$

These numerical relationships are referred to collectively as stoichiometry. The numbers in front of each reactant and product are stoichiometric coefficients. The values of 1 have been shown for clarity but can also be omitted because if no number is shown in a balanced equation, it is assumed to be 1:

$$\text{Tabletop + 4 Table Legs} \rightarrow \text{Table}$$

The stoichiometry in a balanced reaction equation shows the numerical relationships between reactants and products.

An unbalanced reaction equation is useful only for talking about a reaction in a *qualitative* sense. The unbalanced equation (Tabletops + Table Legs → Tables) allows us to describe the components used in making tables. Sometimes that's all we wish to convey, so unbalanced equations have their place.

However, if we want to discuss a reaction *quantitatively* (i.e., if we want to talk about theoretical and percent yield), we must have a balanced reaction equation.

Before you begin any theoretical yield calculation, you must have a balanced reaction equation. Sometimes unbalanced reaction equations are

  DOI: 10.1201/9781003587408-2

presented. Therefore, you must always check to ensure that the equation is balanced, and if it isn't, then you must balance it.

In the previous chapter, we calculated theoretical yield by inspection. In other words, we had diagrams that enabled us to glance at an array of reactants and determine how much product we could make from them. With chemical reactions, this will not be possible. We will need to use calculations to determine theoretical yield. So, let's begin the transition using the balanced reaction equation we've already produced:

$$1 \text{ Tabletop} + 4 \text{ Table Legs} \rightarrow 1 \text{ Table}$$

Imagine a small business that is producing fully assembled tables. Let's say that, on a given day, the business receives a shipment of 875 tabletops and 3,500 table legs. How many tables can the business build that day? Simple calculations using the equation's stoichiometry will enable us to figure this out. We will essentially be performing dimensional analysis, which you may have also learned as the factor-label method or the unit factor method. While those are fancy names, the principle is straightforward: We'll begin with a given quantity of reactant and convert it to the amount of product it could yield. To do this, we'll use two important guidelines:

1. The necessary conversion factor comes from the stoichiometry of the balanced equation.
2. Make sure that you are always canceling old units to obtain new ones.

We'll need a pair of calculations because there are two reactants. First, let's calculate how many tables can be made from the 875 available tabletops.

$$875 \, tabletops \times \frac{1 \, table}{1 \, tabletop} = 875 \, tables$$

Notice a couple of things about this simple calculation. We begin with a given quantity. Then, we look to the balanced equation to determine the stoichiometric ratio (1 tabletop:1 table), which gives us a conversion factor. Finally, we arrange that conversion factor with tabletops in the denominator so that old units cancel to give us new ones. Let's highlight the cancellation of units:

$$875 \, \cancel{tabletops} \times \frac{1 \, table}{1 \, \cancel{tabletop}} = 875 \, tables$$

Notice that the numbers don't cancel. Obviously, 875 and 1 *don't* cancel one another. It is only the unit "tabletops" that cancels. This leaves us with a new unit, "tables."

Now, we need a second calculation to determine how many tables can be made with the available number of table legs.

$$3,500 \; \cancel{table\;legs} \times \frac{1\;table}{4\;\cancel{table\;legs}} = 875\;tables$$

Once again, we begin with a given quantity—this time, the number of table legs. Then, we look to the stoichiometry of the balanced equation to determine a conversion factor (4 table legs:1 table). Finally, we arrange that conversion factor with table legs in the denominator so that the initial units cancel, leaving us with new ones.

In this case, our calculations show us that we have a perfect match! There are enough tabletops to make 875 tables, and happily, there are enough table legs to make 875 tables as well. In this case, there is no limiting reactant because both reactants are consumed simultaneously. You could also say that both reactants are limiting, which would mean the same thing.

Despite how smoothly our first example worked out, we know that reactants won't always be this perfectly matched. Consider a second scenario in which the small business receives a shipment of 875 tabletops and 3,721 table legs. How many tables can the business build on that day? To find the answer, we'll complete our two calculations in the same way but with the new values:

$$875 \; \cancel{tabletops} \times \frac{1\;table}{1\;\cancel{tabletop}} = 875\;tables$$

$$3,721 \; \cancel{table\;legs} \times \frac{1\;table}{4\;\cancel{table\;legs}} = 930.25\;tables$$

This time the table legs are present in excess. In other words, we have more table legs than we need based on the available number of tabletops. There are only enough tabletops to make 875 tables. The tabletops run out first, so they are the limiting reactant, which determines our theoretical yield. The theoretical yield is 875 tables. The fact that we have enough table legs to make 930.25 tables is completely irrelevant. Once the tabletops run out, the remaining table legs are simply extras, and we can't do anything productive with them.

Finally, let's consider a third scenario. The small business receives a shipment of 936 tabletops and 3,500 table legs. How many tables can the business build on that day? To find the answer, we'll complete our two calculations in the same way but with the new values:

$$936 \; \cancel{tabletops} \times \frac{1\;table}{1\;\cancel{tabletop}} = 936\;tables$$

$$3,500 \ \cancel{table \ legs} \times \frac{1 \ table}{4 \ \cancel{table \ legs}} = 875 \ tables$$

In this instance, the tabletops are present in excess. We have more of those than we need. There are only enough table legs to make 875 tables. The table legs run out first, so they are the limiting reactant, which determines the theoretical yield. The theoretical yield is 875 tables. The fact that we have extra tabletops (enough to make 936 tables) is totally irrelevant. Once those table legs run out, the game is over. We can't do anything with the surplus tabletops.

To recap, we've now analyzed three scenarios:

Scenario 1: The number of tabletops and table legs was perfectly matched, giving a theoretical yield of 875 tables.
Scenario 2: The tabletops were the limiting reactant, and the table legs were present in excess. The theoretical yield was still 875 tables.
Scenario 3: The tabletops were present in excess, and the table legs were the limiting reactant. The theoretical yield was 875 tables once again.

These scenarios highlight the fact that we can arrive at the same answer for theoretical yield in different ways. In each case though, we performed one calculation per reactant to see how many products could be made from it. **The smaller amount of product that can be made from given amounts of reactants is always the theoretical yield.**

CHAPTER 2 PRACTICE PROBLEMS

1. Remember our bookshelf example from Chapter 1? Let's use it again. To refresh your memory, the schematic is presented in Figure 2.1

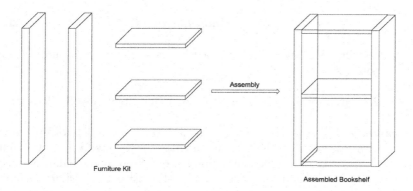

Figure 2.1 Bookshelf assembly from a kit.

(a) Write an *unbalanced* reaction equation that reflects the schematic above.

(b) Now, write a *balanced* reaction equation using the ratios shown in the schematic.

2. We've started a small business building bookshelves of the type shown in Problem #1. For each scenario given below, determine the limiting reactant and calculate the theoretical yield.

(a) We receive a shipment of 1,100 long vertical pieces and 1,650 short horizontal pieces.

(b) We receive a shipment of 3,242 long vertical pieces and 5,002 short horizontal pieces.

(c) We receive a shipment of 753 long vertical pieces and 1,113 short horizontal pieces.

3. Today, our small business receives a shipment of 522 long vertical pieces and 723 short horizontal pieces. We build 176 bookshelves. What is the percent yield for our day's labor?

SOLUTIONS TO CHAPTER 2 PRACTICE PROBLEMS

1. (a) To build a bookshelf, we need long vertical pieces and short horizontal pieces. Translating this into a reaction equation with reactants on the left and products on the right, we have

Long vertical pieces + Short horizontal pieces → Bookshelf

This reaction equation is unbalanced because we've paid no attention to the numerical relationships in our "recipe."

(b) To balance the reaction equation, we must insert numbers before each entry that reflect the schematic:

2 Long vertical pieces + 3 Short horizontal pieces → 1 Bookshelf

Remember that the 1 can be shown for clarity or omitted, as follows:

2 Long vertical pieces + 3 Short horizontal pieces → Bookshelf

2. (a) In this first scenario, when we calculate the number of bookshelves we can make from each reactant, we find that we have a perfect match! Remember to use the conversion factors given by the balanced reaction equation (2 long vertical pieces:1 bookshelf and 3 short horizontal pieces:1 bookshelf). Also, remember to arrange your conversion factors so that the original units cancel, giving you new units.

$$1,100 \; \cancel{long\;vertical\;pieces} \times \frac{1\;bookshelf}{2\;\cancel{long\;vertical\;pieces}} = 550 \; bookshelves$$

$$1,650 \; \cancel{short\;horizontal\;pieces} \times \frac{1\;bookshelf}{3\;\cancel{short\;horizontal\;pieces}} = 550 bookshelves$$

In this case, you can say that neither reagent is limiting or both are limiting. It amounts to the same thing. The theoretical yield is 550 bookshelves.

(b) In this second scenario, our two calculations follow the same strategy but, of course, use the new given values:

$$3,242 \; \cancel{long\;vertical\;pieces} \times \frac{1\;bookshelf}{2\;\cancel{long\;vertical\;pieces}} = 1,621 \; bookshelves$$

$$5,002 \; \cancel{short\;horizontal\;pieces} \times \frac{1\;bookshelf}{3\;\cancel{short\;horizontal\;pieces}}$$
$$\approx 1,667.3 \; bookshelves$$

This time the reactants are not perfectly matched. The long vertical pieces run out first and therefore limit the amount of product we can make, so the long vertical pieces are the limiting reactant. The short horizontal pieces are simply present in excess.

The limiting reactant (long vertical pieces) determines the theoretical yield, which is 1,621 bookshelves.

(c) In the third scenario, our two calculations will look very similar but with the new values in the shipment:

$$753 \; \cancel{long\;vertical\;pieces} \times \frac{1\;bookshelf}{2\;\cancel{long\;vertical\;pieces}} = 376.5 bookshelves$$

$$1,113 \; \cancel{short\;horizontal\;pieces} \times \frac{1\;bookshelf}{3\;\cancel{short\;horizontal\;pieces}}$$
$$= 371 \; bookshelves$$

Once again, the reactants are not perfectly matched. This time though, the short horizontal pieces run out first. They limit the amount of product we can build, so the short horizontal pieces are the limiting reactant. The theoretical yield is always the smaller amount of product that can be made from given amounts of reactants. In this case, that is 371 bookshelves.

3. We must proceed through the determination of limiting reactant and theoretical yield before we can address percent yield. The calculations for determination of limiting reactant and theoretical yield will look much like they did in Problem #2:

$$522 \, \cancel{long\,vertical\,pieces} \times \frac{1\,bookshelf}{2 \, \cancel{long\,vertical\,pieces}} = 261\,bookshelves$$

$$723 \, \cancel{short\,horizontal\,pieces} \times \frac{1\,bookshelf}{3 \, \cancel{short\,horizontal\,pieces}} = 241\,bookshelves$$

These calculations show that the short horizontal pieces run out first and are therefore the limiting reactant. They dictate that the theoretical yield is 241 bookshelves. Our long vertical pieces are present in excess, so the exact number of bookshelves we can make from them is irrelevant.

Recall from Chapter 1 that the percent yield is calculated as follows:

$$Percent\,Yield = \frac{amount\,of\,product\,obtained}{theoretical\,yield} \times 100$$

Inserting specific values for amount of product obtained and theoretical yield shows the percent yield to be as follows:

$$Percent\,Yield = \frac{176\,bookshelves\,obtained}{241\,bookshelves\,it\,was\,possible\,to\,build} \times 100 \approx 73\%$$

Note that the units cancel in a percent yield calculation because, formally, the unit in the numerator and denominator is simply "bookshelves." The extra descriptive text in the calculation above was merely added for clarity.

$$Percent\,Yield = \frac{176 \, \cancel{bookshelves}}{241 \, \cancel{bookshelves}} \times 100 \approx 73\%$$

Chapter 3

Dealing with chemical reactions
The mole

Thus far, our conversation about theoretical yield has dealt with large objects (tables and bookshelves) that we can easily envision because they are part of the macroscopic world. When we shift our focus to the microscopic world, the principles remain exactly the same, but we'll usually be dealing with *much* larger numbers. When you are considering tables, you might think of dozens of them or even hundreds or thousands. However, when you think of molecules, even the smallest samples we can conveniently weigh in the laboratory will contain trillions and trillions of molecules. Therefore, instead of using a unit such as dozen, which might be handy for bookshelves, we utilize the mole. Since we don't encounter this unit in everyday life, it may seem confusing at first. The key to understanding the mole is to realize that it is no different than any other quantity, such as a pair or a dozen.

$$1 \text{ pair} = 2 \text{ things}$$

$$1 \text{ dozen} = 12 \text{ things}$$

$$1 \text{ mole} = 6.022 \times 10^{23} \text{ things}$$

Pairs are convenient when talking about items like socks. Dozens are handy when tallying donuts. Moles are useful when you are discussing *enormous* numbers of things. There would be no point in using the mole when talking about socks or donuts because no one has ever seen that many socks or donuts (remember $6.022 \times 10^{23} = 620{,}000{,}000{,}000{,}000{,}000{,}000{,}000{,}000$, which is a heck of a lot!). Conversely, when we are weighing out even very small samples in the laboratory, talking about pairs or dozens of molecules would be pointless. There are so many molecules in even just 1 milligram that we'd have a gigantic number of pairs or dozens of molecules. However, since the mole is such a big value, it is a convenient way of expressing how many molecules we have in our samples.

DOI: 10.1201/9781003587408-3

The mole is a huge value (6.022×10^{23}), so it is typically useful only when dealing with minuscule objects, such as atoms or molecules.

While you could, in theory, talk about a mole of anything, you'll never see anywhere near that many of most objects. In practice, you'll usually only want to talk about moles of atoms or molecules, which means that the relevant conversion factor is

$$1 \text{ mole} = 6.022 \times 10^{23} \text{ atoms or molecules}$$

How do we know which is relevant—atoms or molecules? That's easy: we'll use "atoms" when we are talking about a sample that consists of only individual atoms, for example, a bar of gold, and we'll use "molecules" when we are considering a sample that consists of molecules instead, such as a dosage of aspirin.

Why does all this matter anyway? What's the point? **The reason the mole matters is that the stoichiometric coefficients in a balanced chemical reaction refer only to molecules or moles.** Consider the balanced chemical reaction for the synthesis of aspirin shown in Figure 3.1.

There is an implied stoichiometric coefficient of 1 in front of each reactant and product. This stoichiometry tells us one of two things. We can say:

> One molecule of salicyclic acid and one molecule of acetic anhydride are needed to make one molecule of aspirin and one molecule of acetic acid.

Or we can say:

> One mole of salicylic acid and one mole of acetic anhydride are needed to make one mole of aspirin and one mole of acetic acid.

However, the stoichiometry *cannot* be used to relate grams, milliliters, or any other quantity of reactants and products. Therefore, when we are in laboratory situations where we are weighing or otherwise dispensing visible quantities of reactants, we'll need to convert those quantities to moles before we can utilize the reaction's stoichiometry.

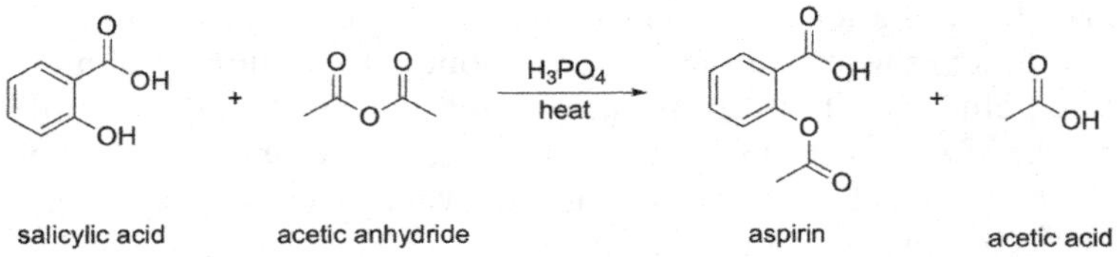

Figure 3.1 Aspirin synthesis.

Let's consider an example, but before we do, two notes about the reaction equation shown in Figure 3.1. First, phosphoric acid (H_3PO_4) is a catalyst. Catalysts are neither created nor destroyed during chemical reactions; they merely accelerate the rate of reaction. Consequently, catalysts are neither reactants nor products and do not need to be considered when determining limiting reagent and calculating theoretical yield. Second, heat is sometimes applied to chemical transformations to make them proceed more quickly. However, it is not something that factors into the calculation of theoretical yield either. We will simply make the assumption that we are heating sufficiently and for long enough that the reaction could proceed to completion.

Now, for our example, imagine that we combine 5.00 grams of salicylic acid with 5.00 grams of acetic anhydride, and as a result, we prepare 4.21 grams of aspirin. What is the limiting reagent? How much aspirin could we prepare (i.e., what is our theoretical yield)? And, what is the percent yield?

We'll need a bit more information to complete our calculations. Namely, we'll need the molar masses of the reactants and product of interest, which are given in Figure 3.2. The molar mass of a substance tells us how many grams (g) of that substance are in a mole (sometimes abbreviated as mol).

Our calculations will have two steps more than the calculations we performed in Chapter 2. In Chapter 2, we immediately used the stoichiometry of the balanced equation. Before we can do that in these cases, we must first convert grams of reactant to moles of reactant using the molar mass. Then we can use the stoichiometry. If we stopped at that point, we'd have moles of product, but since we can't weigh moles directly in the laboratory, it is usually convenient to end with conversion of moles of product to grams of product.

$$5.00 \; \text{g salicylic acid} \times \frac{1 \; \text{mol salicylic acid}}{138.12 \; \text{g salicylic acid}} \times \frac{1 \; \text{mol aspirin}}{1 \; \text{mol salicylic acid}}$$
$$\times \frac{180.16 \; \text{g aspirin}}{1 \; \text{mol aspirin}} \approx 6.52 \; \text{g aspirin}$$

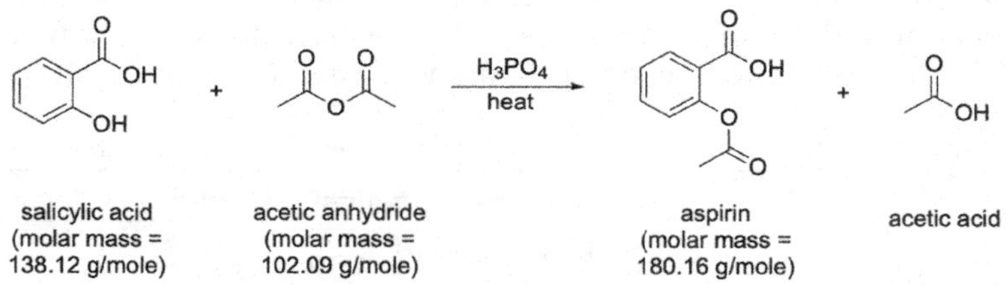

| salicylic acid (molar mass = 138.12 g/mole) | acetic anhydride (molar mass = 102.09 g/mole) | aspirin (molar mass = 180.16 g/mole) | acetic acid |

Figure 3.2 Aspirin synthesis.

$$5.00 \ \cancel{g \ acetic \ anhydride} \times \frac{1 \ \cancel{mol \ acetic \ anhydride}}{102.09 \ \cancel{g \ acetic \ anhydride}}$$

$$\times \frac{1 \ \cancel{mol \ aspirin}}{1 \ \cancel{mol \ acetic \ anhydride}} \times \frac{180.16 \ g \ aspirin}{1 \ \cancel{mol \ aspirin}}$$

$$\approx 8.82 \ g \ aspirin$$

After performing the calculation for each reactant, we see that salicylic acid is the limiting reactant and the theoretical yield is 6.52 grams of aspirin. Lastly, with the theoretical yield in hand, we can now calculate the percent yield based on the amount of product we actually obtained. We obtained a percent yield of 64.6%.

$$Percent \ Yield = \frac{amount \ of \ product \ obtained}{theoretical \ yield} \times 100$$

$$Percent \ Yield = \frac{4.21 \ \cancel{g \ aspirin}}{6.52 \ \cancel{g \ aspirin}} \times 100 \approx 64.6\%$$

As we reflect on this example, there are a few things we should notice to ensure success in future problems:

1. Always write not only the unit (e.g., g or mol) but also the substance. For instance, don't just write "1 mol." Instead, write "1 mol salicylic acid."
2. Although the mole ratio in this problem happened to be 1:1 and therefore does not mathematically affect the outcome of the calculation, do not omit this step. The mole ratio won't always be 1:1, so sometimes it will impact the arithmetic outcome.

To highlight the importance of the two guidelines above, consider their impact on the middle conversion highlighted in the box below. If you had not written the substance names along with the units, it would be difficult to keep track of which substance is in the numerator and which is in the denominator. Although it wouldn't change the outcome in this instance if you got confused (because the ratio is 1:1), it would impact the outcome if the mole ratio were anything other than 1:1 as we'll see in future problems.

$$5.00 \ \cancel{g \ salicylic \ acid} \times \frac{1 \ mol \ \cancel{salicylic \ acid}}{138.12 \ \cancel{g \ salicylic \ acid}} \times \boxed{\frac{1 \ \cancel{mol \ aspirin}}{1 \ \cancel{mol \ salicylic \ acid}}}$$

$$\times \frac{180.16 \ g \ aspirin}{1 \ \cancel{mol \ aspirin}} \approx 6.52 \ g \ aspirin$$

CHAPTER 3 PRACTICE PROBLEMS

1. Consider the reaction of phenylacetylene with bromine shown in Figure 3.3.

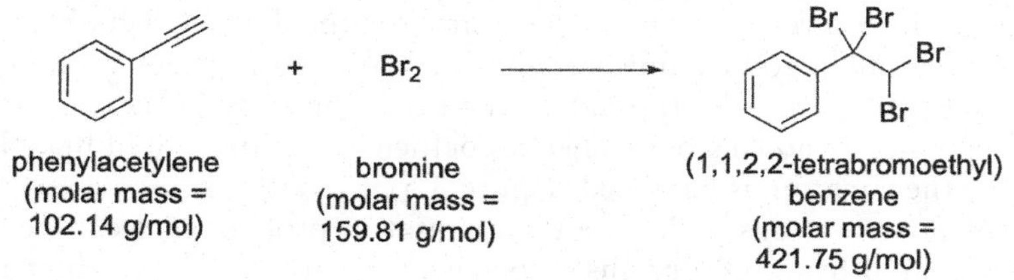
+
Br_2
Br Br
Br
Br
phenylacetylene
(molar mass =
102.14 g/mol)
bromine
(molar mass =
159.81 g/mol)
(1,1,2,2-tetrabromoethyl)
benzene
(molar mass =
421.75 g/mol)

Figure 3.3 Reaction of phenylacetylene with bromine.

(a) This reaction has been provided to you in its *unbalanced* form. Balance the reaction.
(b) Determine the limiting reagent and calculate the theoretical yield of (1,1,2,2-tetrabromoethyl)benzene if we begin with 8.00 grams of phenylacetylene and 20.0 grams of bromine.
(c) What is the percent yield if we isolate 17.4 grams of product?

2. The exhaustive benzylic bromination of toluene is shown in Figure 3.4.

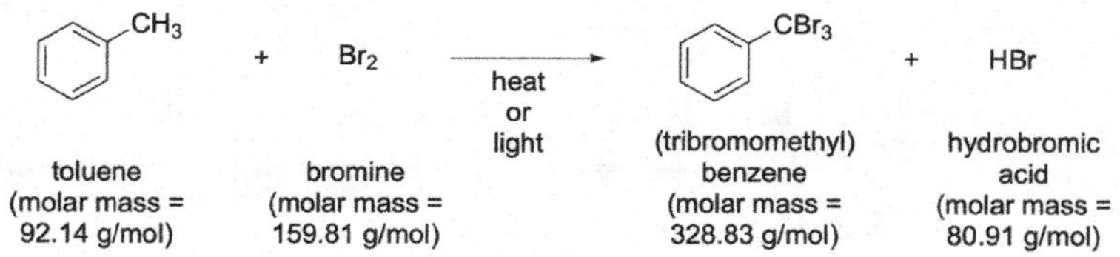
CH_3
+
Br_2
heat
or
light
CBr_3
+
HBr
toluene
(molar mass =
92.14 g/mol)
bromine
(molar mass =
159.81 g/mol)
(tribromomethyl)
benzene
(molar mass =
328.83 g/mol)
hydrobromic
acid
(molar mass =
80.91 g/mol)

Figure 3.4 Exhaustive benzylic bromination of toluene.

(a) If we begin with 6.75 grams of toluene and 28.0 grams of bromine, what is the limiting reactant? What is the theoretical yield of (tribromomethyl)benzene?
(b) What is the percent yield if we obtain 18.5 grams of (tribromomethyl)benzene?

SOLUTIONS TO CHAPTER 3 PRACTICE PROBLEMS

1. (a) We balance equations by inspection. In other words, we examine the number of atoms of each element on both the reactant and the product side to make sure that they are equal. If they are not, we use stoichiometric coefficients *only* to remedy the situation. Note that we

cannot alter the molecular formulas; that would change the identity of the reactants and/or products.

Both sides of the equation contain eight carbon atoms, so the carbons are balanced. Similarly, the hydrogen atoms are balanced because there are six atoms on both the reactant and the product side. However, the bromine atoms are unbalanced. Initially, there are only two bromine atoms on the left side of the equation, while there are four on the right side of the equation. To fix this imbalance, we add a stoichiometric coefficient of 2 in front of Br_2. Now, the equation is balanced (Figure 3.5).

(b) First, we must calculate the amount of product that can be prepared from 8.00 grams of phenylacetylene, which is abbreviated as "pa" in the following calculation. The product's name will be abbreviated as "tbeb."

The mass of phenylacetylene is divided by the molar mass in order to obtain moles of phenylacetylene. Then, we use the mole ratio of phenylacetylene to (1,1,2,2-tetrabromoethyl)benzene, which is 1:1, to convert moles of phenylacetylene to moles of product. Finally, we multiply by the molar mass of the product to obtain the amount of product in grams.

$$8.00 \; \cancel{g \; pa} \times \frac{1 \; \cancel{mol \; pa}}{102.14 \; \cancel{g \; pa}} \times \frac{1 \; \cancel{mol \; tbeb}}{1 \; \cancel{mol \; pa}} \times \frac{421.75 \; g \; tbeb}{1 \; \cancel{mol \; tbeb}} \approx 33.0 \; g \; tbeb$$

Next, we must calculate the amount of product that can be prepared from 20.0 grams of bromine.

$$20.0 \; \cancel{g \; bromine} \times \frac{1 \; \cancel{mol \; bromine}}{159.81 \; \cancel{g \; bromine}} \times \frac{1 \; \cancel{mol \; tbeb}}{2 \; \cancel{mol \; bromine}} \times \frac{421.75 \; g \; tbeb}{1 \; \cancel{mol \; tbeb}}$$
$$\approx 26.4 \; g \; tbeb$$

The reactant that runs out first is bromine. It is consumed only after 26.4 grams of product are produced. Therefore, bromine is

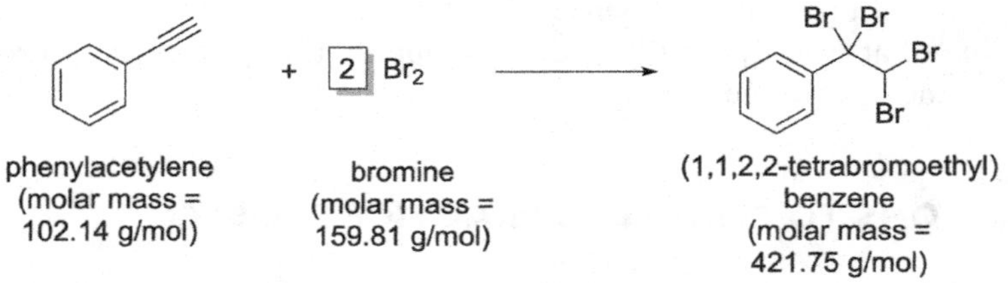

phenylacetylene
(molar mass =
102.14 g/mol)

bromine
(molar mass =
159.81 g/mol)

(1,1,2,2-tetrabromoethyl)
benzene
(molar mass =
421.75 g/mol)

Figure 3.5 Balanced reaction of phenylacetylene with bromine.

the limiting reagent. Remember that the theoretical yield is always the smaller amount of product that is possible to make from the given quantities of reactants. Therefore, the theoretical yield of (1,1,2,2-tetrabromoethyl)benzene is 26.4 grams.

Notice something very important about this example. In this case, the mole ratio was *not* always 1:1. Furthermore, this affected the outcome of our calculations dramatically. If in the second calculation you had omitted the mole ratio altogether or erroneously entered it as 1:1, then you would have obtained an incorrect answer twice as large (≈52.8 grams tbeb). That would have caused you to *misidentify* the limiting reagent as phenylacetylene, and it would have caused you to *incorrectly* state the theoretical yield as 33.0 grams of product. We would have answered the entire question completely incorrectly if we had neglected to carefully consider the mole ratio.

Another item of interest worth pointing out is the importance of always writing not only units but also substances. Below, you'll see the second calculation reproduced but without any of the substance labels. Notice that since the mole ratio (highlighted in the box) has the same unit in the numerator and denominator, it would be very easy to accidentally place the "2" in the numerator rather than in the denominator where it belongs. Doing so would result in an incorrect answer for both limiting reagent and theoretical yield, so it is definitely wise to write both the unit and the substance throughout the entire calculation in order to minimize the chance of such an error.

$$20.0 \; \cancel{g} \times \frac{1 \; \cancel{mol}}{159.81 \; \cancel{g}} \times \boxed{\frac{1 \; \cancel{mol}}{2 \; \cancel{mol}}} \times \frac{421.75 \; g}{1 \; \cancel{mol}} \approx 26.4 \; g$$

(c) Remember that the percent yield is the ratio of product obtained to theoretical yield, multiplied by 100.

$$Percent \; Yield = \frac{amount \; of \; product \; obtained}{theoretical \; yield} \times 100$$

In this case, we obtained 17.4 grams of product, which goes in the numerator. The theoretical yield is 26.4 grams of product, which is placed in the denominator. The units cancel, and we multiply by 100 to obtain approximately a 65.9% yield.

$$Percent \; Yield = \frac{17.4 \; \cancel{g \; product}}{26.4 \; \cancel{g \; product}} \times 100 \approx 65.9\%$$

2. (a) Did you jump right into your calculations without checking to make sure that the reaction equation is balanced? If so, go back and try to balance the reaction before reading the answer below.

It is often not specified whether a reaction equation is balanced or not. It is left to the reader to know that; if they wish to determine limiting reagent and theoretical yield, they will need to balance the reaction before doing so. We balance the equation by inspection. There are seven carbons on each side, so the carbon atoms are balanced. However, there are initially eight hydrogen atoms on the reactant side and only six on the product side. To remedy that imbalance, it is necessary to place a stoichiometric coefficient of 3 in front of HBr (Figure 3.6). Doing so results in having eight hydrogen atoms on both sides of the equation.

Finally, we turn our attention to the bromine atoms. We start with only two bromine atoms on the reactant side. After placing a stoichiometric coefficient of 3 in front of HBr, we have six bromine atoms on the product side. Placing a coefficient of 3 in front of Br_2 leaves us with six bromine atoms on each side of the equation and a fully balanced reaction (Figure 3.7).

With a fully balanced equation, we can now turn our attention to calculations. We have to calculate the amount of product (abbreviated as "tbmb" below) we can make from each of our reactants. Let's begin with the 6.75 grams of toluene. This is converted to moles by dividing by the molar mass. The mole ratio of toluene to (tribromomethyl)benzene is merely 1:1. Lastly, we multiply by the molar mass of (tribromomethyl)benzene to obtain the mass of the product of interest, which is about 24.1 grams.

$$6.75 \; \cancel{g \; toluene} \times \frac{1 \; \cancel{mol \; toluene}}{92.14 \; \cancel{g \; toluene}} \times \frac{1 \; \cancel{mol \; tbmb}}{1 \; \cancel{mol \; toluene}} \times \frac{328.83 \; g \; tbmb}{1 \; \cancel{mol \; tbmb}}$$

$$\approx 24.1 \; g \; tbmb$$

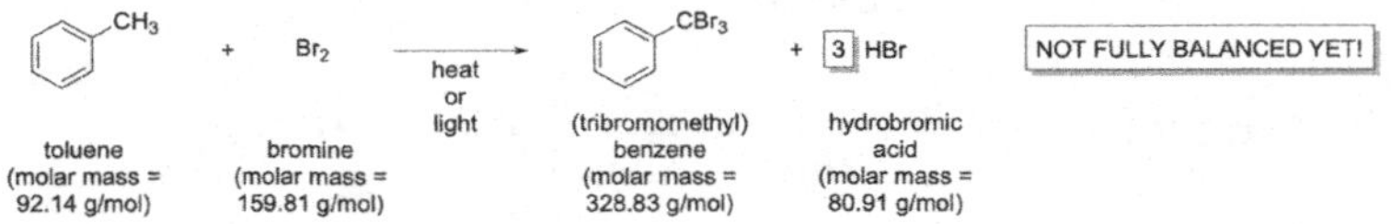

Figure 3.6 Beginning to balance the exhaustive benzylic bromination of toluene.

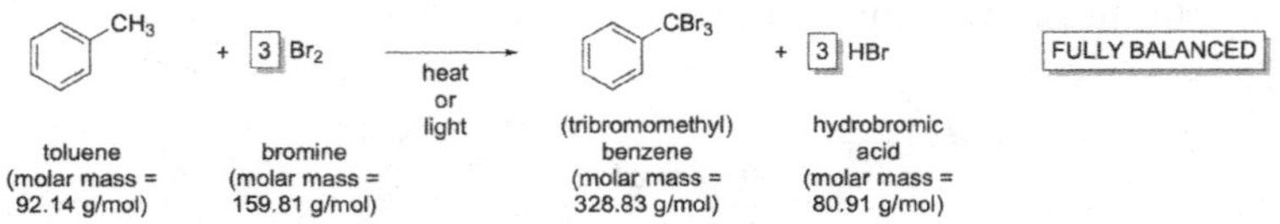

Figure 3.7 Balanced equation for the exhaustive benzylic bromination of toluene.

For our second calculation, division by the molar mass converts the 28.0 grams of bromine to moles of bromine. Then, we must convert moles of bromine to moles of (tribromomethyl)benzene using the mole ratio of 3:1. Be careful here! Don't overlook this step, and make sure that you place the 3 in the denominator where it belongs. Finally, we'll use the molar mass of (tribromomethyl) benzene to convert moles of this product into a mass, giving approximately 19.2 grams.

$$28.0 \; \cancel{g \; bromine} \times \frac{1 \; \cancel{mol \; bromine}}{159.81 \; \cancel{g \; bromine}} \times \frac{1 \; \cancel{mol \; tbmb}}{3 \; \cancel{mol \; bromine}} \times \frac{328.83 \; g \; tbmb}{1 \; \cancel{mol \; tbmb}}$$
$$\approx 19.2 \; g \; tbmb$$

These calculations show that bromine runs out first after only 19.2 grams of product are produced, so it is the limiting reactant. The limiting reactant determines that the theoretical yield is 19.2 grams of (tribromomethyl)benzene.

(b) We can now calculate the percent yield for a reaction that gives 18.5 grams of (tribromomethyl)benzene.

$$Percent \; Yield = \frac{amount \; of \; product \; obtained}{theoretical \; yield} \times 100$$

The 18.5 grams of product obtained occupies the numerator, while the theoretical yield of 19.2 grams goes in the denominator. This gives an impressive percent yield of 96.4%.

$$Percent \; Yield = \frac{18.5 \; \cancel{g \; tbmb}}{19.2 \; \cancel{g \; tbmb}} \times 100 \approx 96.4\%$$

Chapter 4

What if a reactant or product is a neat liquid?

As you can certainly imagine, not all reactants and products will be solids. When a reactant or product is a liquid, it is less convenient to measure its mass. Quantities of liquids are more readily measured as volumes, typically in milliliters for standard laboratory-scale work. How we convert between volume and moles depends on the type of liquid we have. A liquid may be neat (i.e., a pure substance) or a solution (i.e., a solute dissolved in a solvent). In this chapter, we'll address neat liquids, and in the next chapter, we'll discuss solutions.

To reiterate, a neat liquid is a pure substance. For example, a flask of water contains only molecules of H_2O. If we were to dissolve something in that sample of water, it would no longer be a neat liquid. It would become a solution, and we'll consider that situation in the next chapter.

To highlight this concept further, let's revisit an example from the preceding chapter: the synthesis of aspirin from salicylic acid and acetic anhydride (Figure 4.1). It turns out that acetic anhydride is actually a neat liquid. In the previous chapter, we dispensed a certain mass (in grams) of acetic anhydride. Is that feasible? Yes, of course. We can certainly weigh out a quantity of a liquid in an appropriate vessel. However, it is not the most expedient approach. With liquids, we'd prefer to dispense a volume (in milliliters, mL) into a graduated cylinder. So, let's imagine that we do just that. We combine 5.00 grams of salicylic acid with 5.00 mL of acetic anhydride, and as a result, we prepare 4.21 grams aspirin. What is the limiting reagent? How much aspirin could we prepare (i.e., what is our theoretical yield)? And, what is the percent yield?

The calculation for salicylic acid will remain exactly as we saw it in the last chapter. Salicylic acid is a solid, and we dispensed a mass in grams. Therefore, the calculation is straightforward.

$$5.00 \ \cancel{g \ salicylic \ acid} \times \frac{1 \ \cancel{mol \ salicylic \ acid}}{138.12 \ \cancel{g \ salicylic \ acid}} \times \frac{1 \ \cancel{mol \ aspirin}}{1 \ \cancel{mol \ salicylic \ acid}}$$

$$\times \frac{180.16 \ g \ aspirin}{1 \ \cancel{mol \ aspirin}} \approx 6.52 \ g \ aspirin$$

DOI: 10.1201/9781003587408-4

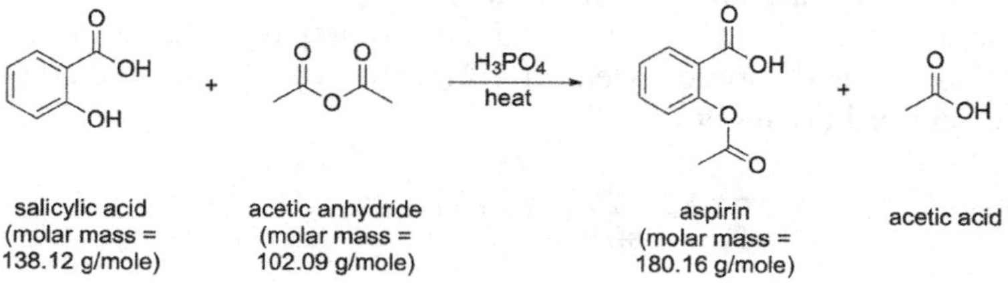

Figure 4.1 Aspirin synthesis.

With acetic anhydride, we'll need just one more step. Density relates the mass of a neat liquid to its volume. The density of acetic anhydride is a physical property, so we look it up in a reference text. The density of acetic anhydride happens to be 1.082 g/mL. We can use this to convert the volume we dispensed into a mass, and from that point on the calculation follows the same strategy we've employed on several occasions. In the calculation that follows, acetic anhydride is abbreviated as "aa."

$$5.00 \; \cancel{mL \; aa} \times \frac{1.082 \; \cancel{g \; aa}}{1 \; \cancel{mL \; aa}} \times \frac{1 \; \cancel{mol \; aa}}{102.09 \; \cancel{g \; aa}} \times \frac{1 \; \cancel{mol \; aspirin}}{1 \; \cancel{mol \; aa}} \times \frac{180.16 \; g \; aspirin}{1 \; \cancel{mol \; aspirin}}$$

$$\approx 9.55 \; g \; aspirin$$

Be sure to guard against two common errors when using volumes of neat liquids:

1. Do *not* simply ignore the fact that your units are milliliter. Sometimes students treat a volume in milliliters as though it were a mass in grams, but notice that if you do that, the units will not cancel. The calculation will be invalid because you'd end up with units that do not make any sense.

$$5.00 \; mL \; aa \times \frac{1 \; \cancel{mol \; aa}}{102.09 \; g \; aa} \times \frac{1 \; \cancel{mol \; aspirin}}{1 \; \cancel{mol \; aa}} \times \frac{180.16 \; g \; aspirin}{1 \; \cancel{mol \; aspirin}}$$

$$\left(Invalid \; calculation; units \; would \; be \frac{(mL \; aa)(g \; aspirin)}{g \; aa} \right)$$

2. Do *not* assume that the density is 1 g/mL. Since water is a commonly encountered neat liquid whose density is about 1 g/mL, sometimes students will assume that all neat liquids have a density of 1 g/mL. This is usually not the case. You cannot know the density of a neat liquid unless you determine it in the laboratory or look it up in a trusted reference. For our purposes in this text, density values will be provided, but when you are doing lab work, you often need to look them up.

Returning to our aspirin synthesis, the limiting reagent is salicylic acid. It runs out first, and there is only enough of it to prepare 6.52 grams of aspirin, which is our theoretical yield. If 4.21 grams of aspirin were prepared, our percent yield is about 64.6%.

$$Percent\ Yield = \frac{4.21\ \cancel{g\ aspirin}}{6.52\ \cancel{g\ aspirin}} \times 100 \approx 64.6\%$$

It is certainly possible that all the reactants and products could be neat liquids. One such example is the preparation of methyl propionate from propionic acid and methanol using catalytic acid and heat (Figure 4.2). Imagine that we start with 5.00 mL of propionic acid and 7.50 mL of methanol and that our synthesis yields 3.67 mL of methyl propionate. We'd want to know what the theoretical yield of methyl propionate is so that we can determine the percent yield for this transformation.

Since both reactants are neat liquids, we'll need to use their densities to convert the volumes in milliliters to masses in grams. Since the product is also a neat liquid, we'll want our last step to convert the mass of product in grams to the volume of product in milliliters. In the calculations that follow, propionic acid is abbreviated as "pa," methanol is abbreviated as "m," and methyl propionate is abbreviated as "mp."

$$5.00\ \cancel{mL\ pa} \times \frac{0.988\ \cancel{g\ pa}}{1\ \cancel{mL\ pa}} \times \frac{1\ \cancel{mol\ pa}}{74.08\ \cancel{g\ pa}} \times \frac{1\ \cancel{mol\ mp}}{1\ \cancel{mol\ pa}} \times \frac{88.11\ \cancel{g\ mp}}{1\ \cancel{mol\ mp}}$$
$$\times \frac{1\ mL\ mp}{0.915\ \cancel{g\ mp}} \approx 6.42\ mL\ methyl\ propionate$$

$$7.50\ \cancel{mL\ m} \times \frac{0.792\ \cancel{g\ m}}{1\ \cancel{mL\ m}} \times \frac{1\ \cancel{mol\ m}}{32.04\ \cancel{g\ m}} \times \frac{1\ \cancel{mol\ mp}}{1\ \cancel{mol\ m}} \times \frac{88.11\ \cancel{g\ mp}}{1\ \cancel{mol\ mp}}$$
$$\times \frac{1\ mL\ mp}{0.915\ \cancel{g\ mp}} \approx 17.9\ mL\ methyl\ propionate$$

In this case, it is the propionic acid that runs out first, making it the limiting reagent and the theoretical yield 6.42 mL of methyl propionate. If we

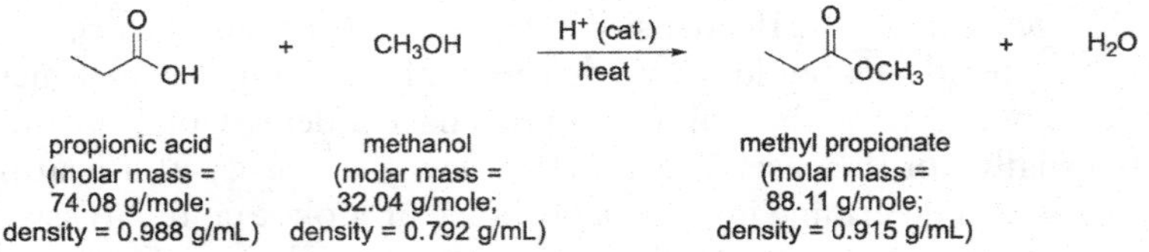

propionic acid   methanol   methyl propionate
(molar mass =   (molar mass =   (molar mass =
74.08 g/mole;   32.04 g/mole;   88.11 g/mole;
density = 0.988 g/mL)   density = 0.792 g/mL)   density = 0.915 g/mL)

Figure 4.2 Methyl propionate synthesis.

prepared 3.67 mL of this product, then the percent yield would be roughly 57.2%.

$$Percent\ Yield = \frac{amount\ of\ product\ obtained}{theoretical\ yield} \times 100$$

$$Percent\ Yield = \frac{3.67\ \cancel{mL\ methyl\ propionate}}{6.42\ \cancel{mL\ methyl\ propionate}} \times 100 \approx 57.2\%$$

Notice that it doesn't matter what the units are in the percent yield calculation (grams versus milliliters); as long as they are the same in both the numerator and denominator, they will cancel.

CHAPTER 4 PRACTICE PROBLEMS

1. Let's revisit a reaction from the Chapter 3 practice problems: the addition of bromine to phenylacetylene (Figure 4.3).
   (a) Determine the limiting reagent and calculate the theoretical yield of (1,1,2,2,-tetrabromoethyl)benzene if we begin with 5.00 mL of phenylacetylene and 4.00 mL of bromine.
   (b) What is the percent yield if we isolate 10.1 grams of product?

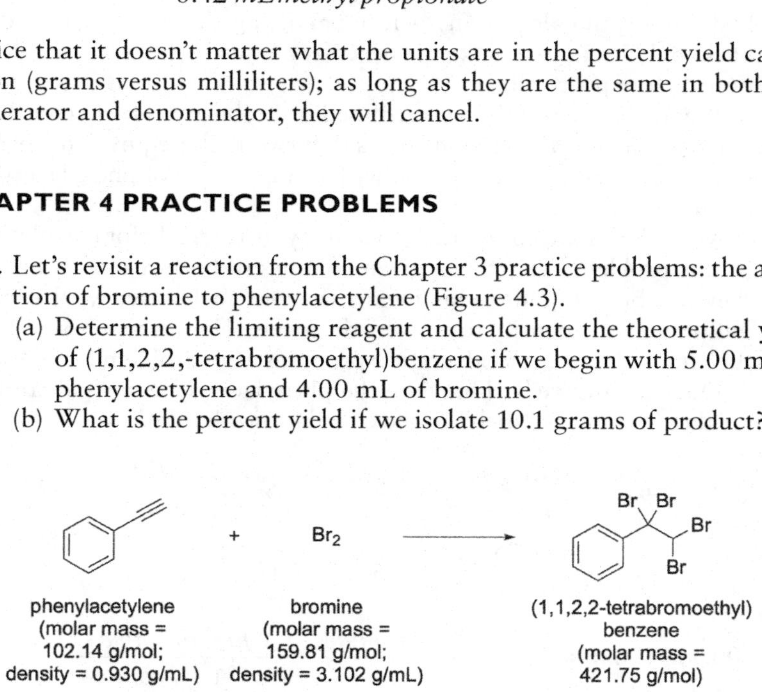
+
Br_2
Br Br
Br
Br
phenylacetylene (molar mass = 102.14 g/mol; density = 0.930 g/mL)
bromine (molar mass = 159.81 g/mol; density = 3.102 g/mL)
(1,1,2,2-tetrabromoethyl) benzene (molar mass = 421.75 g/mol)

Figure 4.3 Reaction of phenylacetylene with bromine.

2. Isoamyl acetate is an ester with the scent of banana. It can be prepared through the following reaction (Figure 4.4).

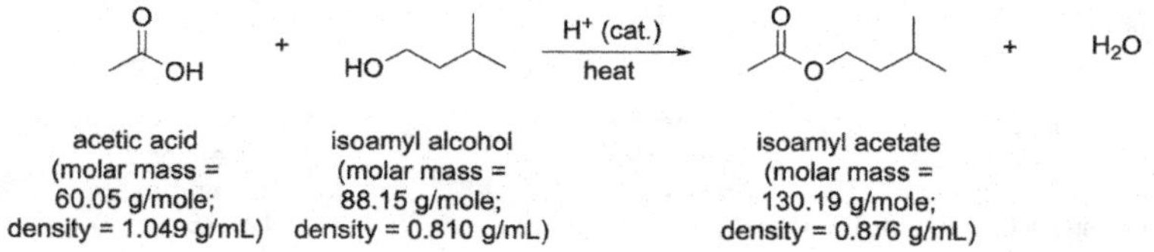
O
OH
+
HO
H^+ (cat.)
heat
O
O
+
H_2O
acetic acid (molar mass = 60.05 g/mole; density = 1.049 g/mL)
isoamyl alcohol (molar mass = 88.15 g/mole; density = 0.810 g/mL)
isoamyl acetate (molar mass = 130.19 g/mole; density = 0.876 g/mL)

Figure 4.4 Isoamyl acetate synthesis.

   (a) Determine the limiting reagent and calculate the theoretical yield of isoamyl acetate if we begin with 10.50 mL of acetic acid and 20.00 mL of isoamyl alcohol.

   (b) What is the percent yield if we isolate 12.36 mL of isoamyl acetate?

SOLUTIONS TO CHAPTER 4 PRACTICE PROBLEMS

1. Did you begin calculations before balancing the equation? If so, go back, balance it, and make the necessary adjustments to your calculations before proceeding.

    Remember that, as we determined in Problem 1 of Chapter 3, the balanced chemical equation shows that two molar equivalents of bromine are necessary for reaction with a single mole of phenylacetylene (Figure 4.5).

    With the balanced equation in hand, you are ready for calculations. In this problem, we've been given volumes of both reactants because they are both liquids. This means that, in each case, we'll need to multiply by the density to convert the given volumes to masses. Then, the remainder of each calculation proceeds as in Problem 1 of Chapter 3. Once again, we'll abbreviate the phenylacetylene as "pa" and the product's name as "tbeb" in the calculations that follow.

$$5.00 \; mL \, pa \times \frac{0.930 \; g \, pa}{1 \; mL \, pa} \times \frac{1 \; mol \, pa}{102.14 \; g \, pa} \times \frac{1 \; mol \, tbeb}{1 \; mol \, pa}$$

$$\times \frac{421.75 \; g \; tbeb}{1 \; mol \, tbeb} \approx 19.2 \; g \; tbeb$$

$$4.00 \; mL \, bromine \times \frac{3.102 \; g \, bromine}{1 \; mL \, bromine} \times \frac{1 \; mol \, bromine}{159.81 \; g \, bromine}$$

$$\times \frac{1 \; mol \, tbeb}{2 \; mol \, bromine} \times \frac{421.75 \; g \; tbeb}{1 \; mol \, tbeb} \approx 16.4 \; g \; tbeb$$

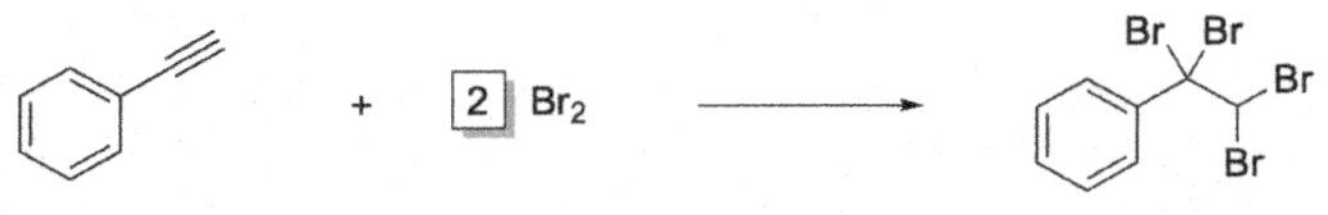

phenylacetylene	bromine	(1,1,2,2-tetrabromoethyl)
(molar mass =	(molar mass =	benzene
102.14 g/mol;	159.81 g/mol;	(molar mass =
density = 0.930 g/mL)	density = 3.102 g/mL)	421.75 g/mol)

Figure 4.5 Balanced reaction of phenylacetylene with bromine.

Bromine is the reactant that runs out first, so it limits the amount of product that can be made to about 16.4 grams. Remember the importance of the mole ratio. In the bromine calculation, if you had not divided by 2 for the mole ratio (2 moles bromine:1 mole product), then you would have erroneously assigned phenylacetylene as the limiting reactant, which would have also resulted in an incorrect theoretical yield.

(b) If we obtained 10.1 grams of product, the percent yield would be roughly 61.6%.

$$Percent\ Yield = \frac{amount\ of\ product\ obtained}{theoretical\ yield} \times 100$$

$$Percent\ Yield = \frac{10.1\ \cancel{g\ tbcb}}{16.4\ \cancel{g\ tbcb}} \times 100 \approx 61.6\%$$

2. The chemical equation for the preparation of isoamyl acetate from acetic acid and isoamyl alcohol is balanced as given, so we need not add any stoichiometric coefficients to it. They are all implied values of 1 (Figure 4.6).

(a) In this case, all of the substances (including the products) are liquids at room temperature, so we've been given volumes exclusively. This means that we'll need to use the appropriate densities at the beginning and end of each calculation, as shown below where acetic acid is abbreviated as "aa," isoamyl alcohol is abbreviated as "ia," and isoamyl acetate is abbreviated as "p" for product.

$$10.50\ \cancel{mL\ aa} \times \frac{1.049\ \cancel{g\ aa}}{1\ \cancel{mL\ aa}} \times \frac{1\ \cancel{mol\ aa}}{60.05\ \cancel{g\ aa}} \times \frac{1\ \cancel{mol\ p}}{1\ \cancel{mol\ aa}} \times \frac{130.19\ \cancel{g\ p}}{1\ \cancel{mol\ p}}$$

$$\times \frac{1\ mL\ p}{0.876\ \cancel{g\ p}} \approx 27.26\ mL\ isoamyl\ acetate\ product$$

$$20.00\ \cancel{mL\ ia} \times \frac{0.810\ \cancel{g\ ia}}{1\ \cancel{mL\ ia}} \times \frac{1\ \cancel{mol\ ia}}{88.15\ \cancel{g\ ia}} \times \frac{1\ \cancel{mol\ p}}{1\ \cancel{mol\ ia}} \times \frac{130.19\ \cancel{g\ p}}{1\ \cancel{mol\ p}}$$

$$\times \frac{1\ mL\ p}{0.876\ \cancel{g\ p}} \approx 27.31\ mL\ isoamyl\ acetate\ product$$

acetic acid	isoamyl alcohol	isoamyl acetate
(molar mass = 60.05 g/mole; density = 1.049 g/mL)	(molar mass = 88.15 g/mole; density = 0.810 g/mL)	(molar mass = 130.19 g/mole; density = 0.876 g/mL)

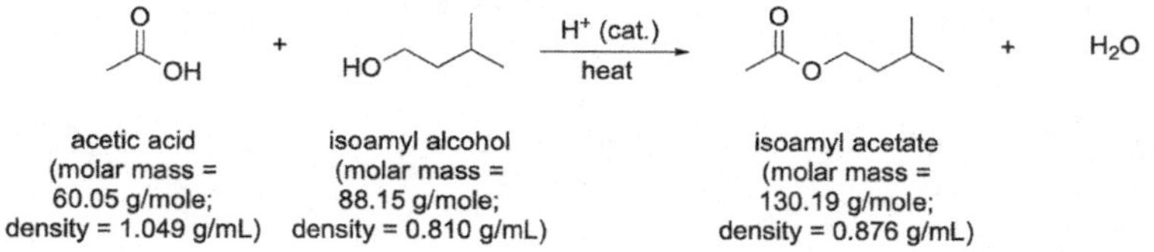

Figure 4.6 Balanced isoamyl acetate synthesis.

In this scenario, the two reactants are nearly perfectly matched, but it is the acetic acid that runs out just a bit before its partner, limiting the theoretical yield to 27.26 mL of isoamyl acetate.

(b) If 12.36 mL of isoamyl acetate are generated, the percent yield is about 45.34%.

$$Percent\ Yield = \frac{amount\ of\ product\ obtained}{theoretical\ yield} \times 100$$

$$Percent\ Yield = \frac{12.36\ \cancel{mL\ isoamyl\ acetate}}{27.26\ \cancel{mL\ isoamyl\ acetate}} \times 100 \approx 45.34\%$$

Chapter 5

What if a reactant is a solution?

In the previous chapter, we noted that a liquid may be neat (i.e., a pure substance) or a solution (i.e., a solute dissolved in a solvent). We've also addressed how to handle calculations for a reactant and/or a product that is a neat liquid, but we have yet to address solutions. That will be the subject of this chapter.

A solution contains a solute (i.e., the substance that is dissolved) in a solvent (i.e., the substance that dissolves the solute). A simple example would be sugar water, where sugar is the solute and water is the solvent. With solutions, concentration must be specified. This tells you how much solute is present per unit of solution volume. You know about concentration intuitively. If only a little sugar is dissolved in a pitcher of water, the sugar water is not very concentrated, and it won't be that sweet. If however you dissolve a large amount of sugar in a pitcher of water, the concentration will be high, and the sweetness will be pronounced. This addresses concentration from a qualitative perspective, but we can also quantify concentration.

A common measure of concentration is molarity (M). Molarity is the moles of solute per liter of solution.

$$Molarity(M) = \frac{moles\,solute}{liter(L)\,of\,solution}$$

When a reactant is delivered as a solution, it is the molarity that is used to convert volume to moles. Let's highlight this through an example. Consider the reaction of cyclohexene with a solution of bromine in dichloromethane (CH_2Cl_2), as shown in Figure 5.1. What is the limiting reactant and theoretical yield if we combine 2.00 mL of cyclohexene with 20.00 mL of a 1.00 M bromine solution? What is the percent yield if 1.48 mL of product are produced?

In this balanced equation, we face two different types of liquids. Cyclohexene is a neat liquid, so we'll use its density to convert volume to mass, similar to the situations we encountered in the preceding chapter. The product is also a neat liquid, so we'll employ its density to convert mass

DOI: 10.1201/9781003587408-5

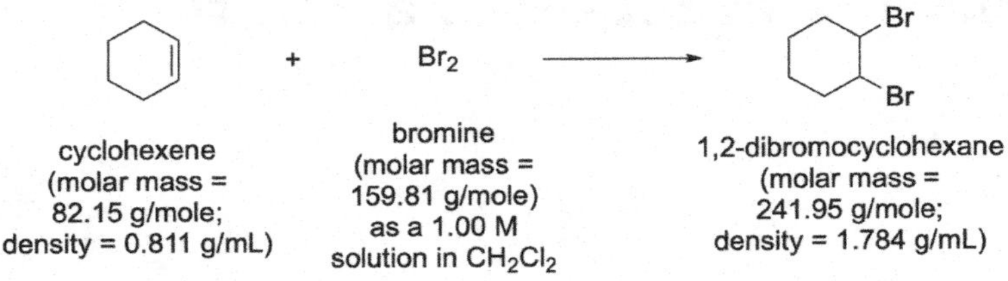

cyclohexene
(molar mass =
82.15 g/mole;
density = 0.811 g/mL)

bromine
(molar mass =
159.81 g/mole)
as a 1.00 M
solution in CH_2Cl_2

1,2-dibromocyclohexane
(molar mass =
241.95 g/mole;
density = 1.784 g/mL)

Figure 5.1 Bromination of cyclohexene.

to volume in the final step of our computation. In the calculation below, cyclohexene and the dibromocyclohexane product are abbreviated as "c" and "dbc," respectively.

$$2.00 \; mL \, c \times \frac{0.811 \; g \, c}{1 \; mL \, c} \times \frac{1 \; mole \, c}{82.15 \; g \, c} \times \frac{1 \; mol \, dbc}{1 \; mole \, c} \times \frac{241.95 \; g \, dbc}{1 \; mol \, dbc}$$

$$\times \frac{1 \; mL \; dbc}{1.784 \; g \, dbc} \approx 2.68 \; mL \; dbc$$

On the other hand, the bromine reagent is delivered as a solution in dichloromethane (CH_2Cl_2), so our approach will differ only slightly in that we will use the concentration of this solution, which was provided a 1.00 M, to convert the volume of solution to the moles of solute (Br_2). Since molarity is moles of solute per *liter* of solution, we must first convert the volume given in milliliters to liters.

$$20.00 \; mL \, Br_2 \, solution \times \frac{1 \; L}{1000 \; mL} \times \frac{1.00 \; mol \, Br_2}{1 \; L \, Br_2 \, solution} \times \frac{1 \; mol \, dbc}{1 \; mol \, Br_2}$$

$$\times \frac{241.95 \; g \, dbc}{1 \; mol \, dbc} \times \frac{1 \; mL \; dbc}{1.784 \; g \, dbc} \approx 2.71 \; mL \; dbc$$

Notice that, in this instance, there is no need for the molar mass of bromine because the solution volume is converted directly to moles using molarity.

Our calculations show that cyclohexene (the limiting reactant) is consumed first after 2.68 mL of 1,2-dibromocyclohexane (the theoretical yield) are generated. If we actually produce 1.48 mL of product, then the percent yield is about 55.2%.

$$Percent \; Yield = \frac{amount \, of \, product \, obtained}{theoretical \, yield} \times 100$$

$$= \frac{1.48 \; mL \, dibromocyclohexane}{2.68 \; mL \, dibromocyclohexane} \times 100 \approx 55.2\%$$

CHAPTER 5 PRACTICE PROBLEMS

1. Saponification is the basic hydrolysis of an ester. The saponification of methyl benzoate is shown in Figure 5.2.
   (a) What is the theoretical yield of sodium benzoate if 10.00 mL of 1.00 M methyl benzoate in THF (tetrahydrofuran) are combined with 5.00 mL of 3.00 M sodium hydroxide in water?
   (b) What is the percent yield if the mass of sodium benzoate obtained is 1.13 g?

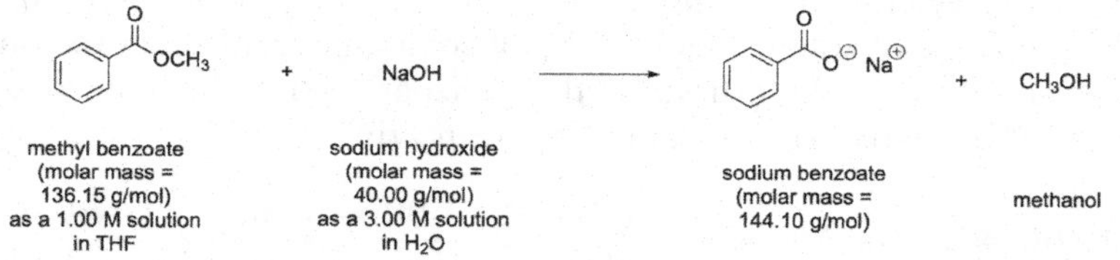

methyl benzoate	sodium hydroxide	sodium benzoate	methanol
(molar mass =	(molar mass =	(molar mass =	
136.15 g/mol)	40.00 g/mol)	144.10 g/mol)	
as a 1.00 M solution	as a 3.00 M solution		
in THF	in H₂O		

Figure 5.2 Saponification of methyl benzoate.

2. Let's revisit the exhaustive benzylic bromination of toluene that we first encountered in Problem 2 of Chapter 3 (Figure 5.3).
   (a) What is the theoretical yield of (tribromomethyl)benzene if we begin with 5.50 mL of toluene and 70.00 mL of 2.25 M bromine solution?
   (b) What is the percent yield if 12.8 grams of (tribromomethyl)benzene are isolated?

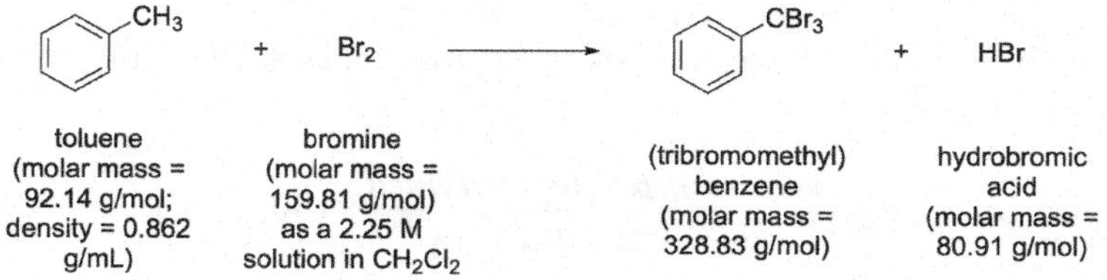

toluene	bromine	(tribromomethyl)	hydrobromic
(molar mass =	(molar mass =	benzene	acid
92.14 g/mol;	159.81 g/mol)	(molar mass =	(molar mass =
density = 0.862	as a 2.25 M	328.83 g/mol)	80.91 g/mol)
g/mL)	solution in CH₂Cl₂		

Figure 5.3 Exhaustive benzylic bromination of toluene.

SOLUTIONS TO CHAPTER 5 PRACTICE PROBLEMS

1. (a) This reaction equation is balanced as given, which simplifies matters (Figure 5.4).

   We find that both reactants are delivered as solutions. The solvents differ, but from the perspective of our theoretical yield

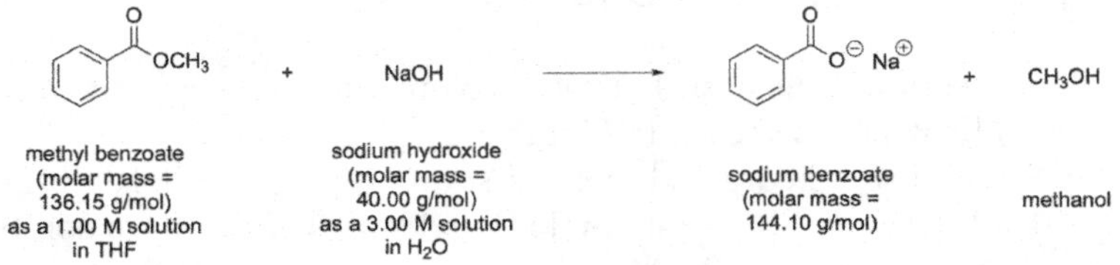

Figure 5.4 Balanced saponification of methyl benzoate.

calculations, the identities of the solvents are irrelevant. Both of our calculations will use the corresponding molarity to convert volume of solution into moles of solute, but remember to convert milliliters to liters first. Methyl benzoate and sodium benzoate are abbreviated as "mb" and "sb," respectively.

$$10.00 \text{ mL mb solution} \times \frac{1 \text{ L}}{1000 \text{ mL}} \times \frac{1.00 \text{ mol mb}}{1 \text{ L mb solution}} \times \frac{1 \text{ mol sb}}{1 \text{ mol mb}}$$
$$\times \frac{144.10 \text{ g sb}}{1 \text{ mol sb}} \approx 1.44 \text{ g sb}$$

$$5.00 \text{ mL NaOH solution} \times \frac{1 \text{ L}}{1000 \text{ mL}} \times \frac{3.00 \text{ mol NaOH}}{1 \text{ L NaOH solution}}$$
$$\times \frac{1 \text{ mol sb}}{1 \text{ mol NaOH}} \times \frac{144.10 \text{ g sb}}{1 \text{ mol sb}} \approx 2.16 \text{ g sb}$$

Our calculations reveal the limiting reactant to be methyl benzoate because it is consumed first, after only 1.44 grams sodium benzoate (the theoretical yield) are formed.

(b) If we make 1.13 grams sodium benzoate, then the percent yield is about 78.5%.

$$\text{Percent Yield} = \frac{\text{amount of product obtained}}{\text{theoretical yield}} \times 100$$
$$= \frac{1.13 \text{ g sodium benzoate}}{1.44 \text{ g sodium benzoate}} \times 100 \approx 78.5\%$$

2. Did you begin to perform calculations without balancing the equation first? If so, go back and balance the equation. Then, modify your calculations accordingly before proceeding.

(a) As we saw in Problem 2 of Chapter 3, the fully balanced equation has stoichiometric coefficients of 3 in front of bromine and HBr (Figure 5.5).

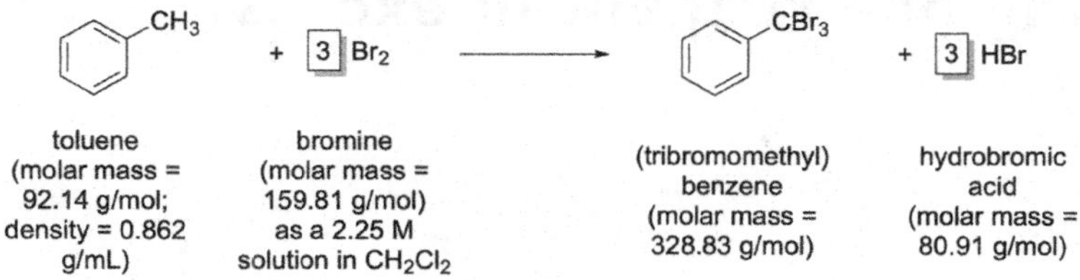

toluene
(molar mass =
92.14 g/mol;
density = 0.862
g/mL)

bromine
(molar mass =
159.81 g/mol)
as a 2.25 M
solution in CH_2Cl_2

(tribromomethyl)
benzene
(molar mass =
328.83 g/mol)

hydrobromic
acid
(molar mass =
80.91 g/mol)

Figure 5.5 Balanced equation for the exhaustive benzylic bromination of toluene.

With the balanced equation in hand, we may continue with our calculations. Notice that our reactants are different types of liquids. Toluene is a neat liquid, so we'll use density to convert the volume provided to mass. The product of interest is abbreviated as "tbmb."

$$5.50 \text{ mL toluene} \times \frac{0.862 \text{ g toluene}}{1 \text{ mL toluene}} \times \frac{1 \text{ mol toluene}}{92.14 \text{ g toluene}}$$
$$\times \frac{1 \text{ mol tbmb}}{1 \text{ mol toluene}} \times \frac{328.83 \text{ g tbmb}}{1 \text{ mol tbmb}} \approx 16.9 \text{ g tbmb}$$

However, the bromine solution will require a different approach. We'll need to use the molarity of this solution to convert volume to moles. Don't forget to (a) convert milliliters to liters before using the molarity and (b) use the correct mole ratio (3 moles bromine:1 mole (tribromomethyl)benzene).

$$70.00 \text{ mL Br}_2 \text{ solution} \times \frac{1 \text{ L}}{1000 \text{ mL}} \times \frac{2.25 \text{ mol Br}_2}{1 \text{ L Br}_2 \text{ solution}} \times \frac{1 \text{ mol tbmb}}{3 \text{ mol Br}_2}$$
$$\times \frac{328.83 \text{ g tbmb}}{1 \text{ mol tbmb}} \approx 17.3 \text{ g tbmb}$$

Our calculations show that toluene runs out first and is therefore the limiting reactant. The theoretical yield is 16.9 grams (tribromomethyl)benzene.

(b) If the actual yield of (tribromomethyl)benzene is 12.8 grams, then the percent yield is roughly 75.7%.

$$Percent\,Yield = \frac{amount\,of\,product\,obtained}{theoretical\,yield} \times 100$$
$$= \frac{12.8 \text{ g tbmb}}{16.9 \text{ g tbmb}} \times 100 \approx 75.7\%$$

Chapter 6

Using one reagent in excess

Although all starting materials in a chemical reaction are formally reactants, sometimes one is more valuable than the other and is considered *the* reactant. The other reactant, whose real role is to modify the reactant of interest, can be referred to as the reagent. In such cases, it is not uncommon to use the reagent in excess so that all of the valuable reactant is transformed into product. When a reagent is used in excess, we are saying that we have more of it than necessary, and it will *not* therefore be the limiting reactant.

Consider the synthesis of methyl salicylate, an ester with a minty aroma, via Fischer esterification, as shown in Figure 6.1. Methanol is a reactant, but it is a simple substance that is inexpensive and widely available. Salicylic acid is a slightly more exotic molecule, so it is the reactant of interest. Methanol is the reagent used to modify our more valuable salicylic acid, so it might be used in excess.

Imagine that we heat 5.00 grams of salicylic acid with an excess of methanol and catalytic acid. We produce 3.98 mL of methyl salicylate. What is the percent yield?

In order to determine percent yield, we must first calculate the theoretical yield. We know that a catalyst is neither created nor destroyed in a chemical reaction, so it can be ignored when figuring theoretical yield. By saying that the methanol is used in excess, we automatically know that there is more methanol than we need to convert all of the salicylic acid into methyl

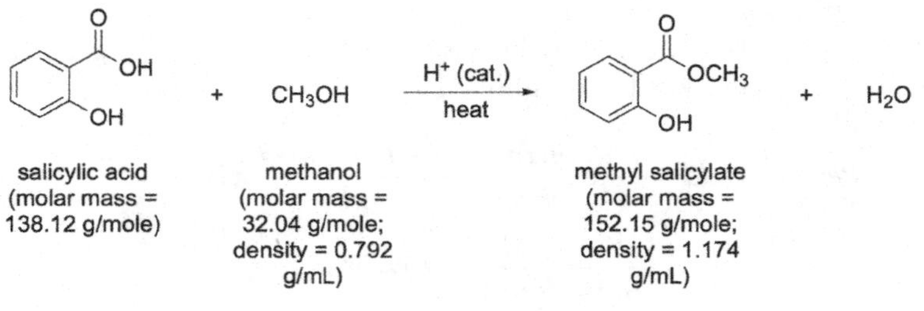

| salicylic acid (molar mass = 138.12 g/mole) | methanol (molar mass = 32.04 g/mole; density = 0.792 g/mL) | methyl salicylate (molar mass = 152.15 g/mole; density = 1.174 g/mL) |

Figure 6.1 Methyl salicylate synthesis.

DOI: 10.1201/9781003587408-6

salicylate. This simplifies our job because we don't have to consider the methanol when calculating theoretical yield either. The phrasing of this question indirectly informs us that the salicylic acid is the limiting reactant. It alone will determined the theoretical yield, which means that only one calculation is needed, even though there are formally two reactants.

We'll begin with the mass of salicylic acid (abbreviated as "sa" below) and calculate the volume of methyl salicylate (abbreviated as "ms") produced. Since the chemical equation is balanced as is, the mole ratio is simply 1:1.

$$5.00 \; \cancel{g \; sa} \times \frac{1 \; \cancel{mol \; sa}}{138.12 \; \cancel{g \; sa}} \times \frac{1 \; \cancel{mol \; ms}}{1 \; \cancel{mol \; sa}} \times \frac{152.15 \; \cancel{g \; ms}}{1 \; \cancel{mol \; ms}} \times \frac{1 \; mL \; ms}{1.174 \; \cancel{g \; ms}}$$

$$\approx 4.69 \; mL \; methyl \; salicylate$$

Now that we know the theoretical yield, we are able to ascertain the percent yield.

$$Percent \; Yield = \frac{amount \; of \; product \; obtained}{theoretical \; yield} \times 100$$

$$= \frac{3.98 \; \cancel{mL \; ms}}{4.69 \; \cancel{mL \; ms}} \times 100 \approx 84.9\%$$

When a question states that a reagent is used in excess, it is a simplification of the problem. That statement allows us to focus only on the reactant of interest when determining theoretical yield.

It is also worth noting that, in such situations, the reagent is commonly shown above or below the reaction arrow, as seen in Figure 6.2. The use of excess may or may not be stipulated in the diagram. We *cannot* assume a reagent is used in excess simply because it is drawn above or below the reaction arrow though. If we are not told that it is used in excess, then we'd have to consider it as a contender for the limiting reagent.

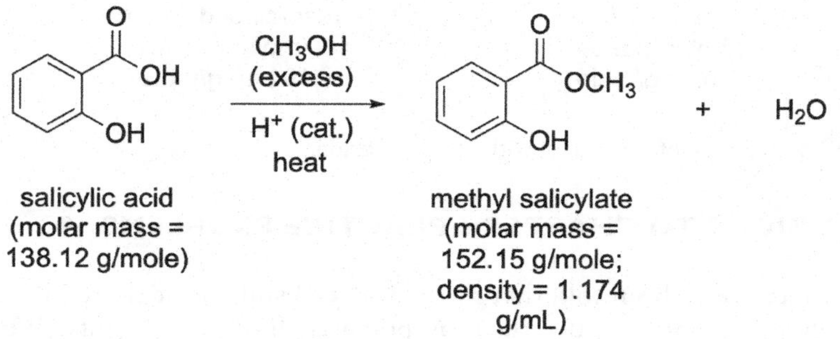

Figure 6.2 Methyl salicylate synthesis shown in an alternative fashion.

CHAPTER 6 PRACTICE PROBLEMS

1. Methyl benzoate (3.50 mL) is treated with an excess of nitric and sulfuric acid (HNO_3 and H_2SO_4, respectively) as shown in Figure 6.3. As a result, 3.75 grams of methyl *meta*-nitrobenzoate are produced. What is the percent yield for this reaction?

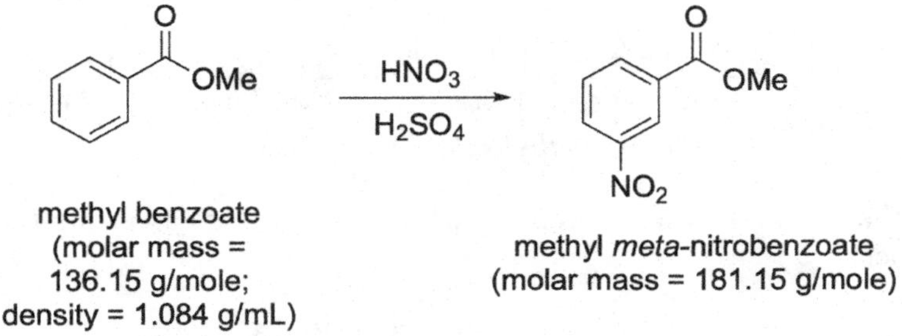

OMe
HNO3
H2SO4
OMe
NO2
methyl benzoate
(molar mass =
136.15 g/mole;
density = 1.084 g/mL)
methyl meta-nitrobenzoate
(molar mass = 181.15 g/mole)

Figure 6.3 Nitration of methyl benzoate.

2. Resorcinol (3.75 grams) is treated with an excess of acetyl chloride to yield 4.11 grams of resorcinol diacetate (Figure 6.4). What is the percent yield?

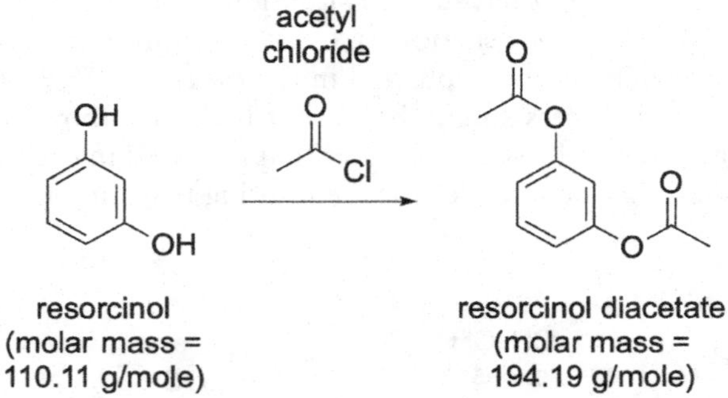

acetyl
chloride
OH
OH
Cl
resorcinol
(molar mass =
110.11 g/mole)
resorcinol diacetate
(molar mass =
194.19 g/mole)

Figure 6.4 Reaction of resorcinol with acetyl chloride.

SOLUTIONS TO CHAPTER 6 PRACTICE PROBLEMS

1. Since we've been told that nitric acid and sulfuric acid are in excess, we need not consider them as potential limiting reagents. We can focus on the methyl benzoate, which must be the limiting reactant.

Furthermore, this reaction is balanced as given with respect to the methyl benzoate and methyl *meta*-nitrobenzoate. Let's perform the calculation of theoretical yield. Note that methyl benzoate and methyl *meta*-nitrobenzoate will be abbreviated as "mb" and "mmnb," respectively.

$$3.50 \; \cancel{mL \; mb} \times \frac{1.084 \; \cancel{g \; mb}}{1 \; \cancel{mL \; mb}} \times \frac{1 \; \cancel{mole \; mb}}{136.15 \; \cancel{g \; mb}} \times \frac{1 \; \cancel{mole \; mmnb}}{1 \; \cancel{mole \; mb}}$$

$$\times \frac{181.15 \; g \; mmnb}{1 \; \cancel{mole \; mmnb}} \approx 5.05 \; g \; methyl \; meta-nitrobenzoate$$

Now that we have the theoretical yield, we are able to calculate percent yield.

$$Percent \; Yield = \frac{amount \; of \; product \; obtained}{theoretical \; yield} \times 100$$

$$= \frac{3.75 \; \cancel{g \; mmnb}}{5.05 \; \cancel{g \; mmnb}} \times 100 \approx 74.3\%$$

2. This equation was not balanced as given. However, when we do balance the equation by placing a 2 in front of acetyl chloride and determining that two molar equivalents of HCl are produced, we see that the stoichiometric coefficients will have no impact on the calculation (Figure 6.5). Since we were told that acetyl chloride is used in excess, it is definitely *not* the limiting reagent. Resorcinol must be the limiting reactant, and it is in a 1:1 mole ratio with the resorcinol diacetate product.

We are now able to complete the calculation of theoretical yield with confidence that we have the correct mole ratio. Note that resorcinol diacetate is abbreviated as "rd" below.

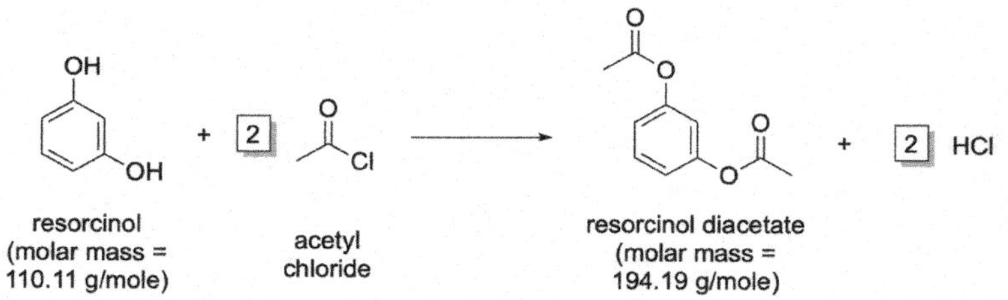

resorcinol
(molar mass =
110.11 g/mole)

acetyl
chloride

resorcinol diacetate
(molar mass =
194.19 g/mole)

Figure 6.5 Balanced reaction of resorcinol with acetyl chloride.

$$3.75 \; \cancel{g \; resorcinol} \times \frac{1 \; \cancel{mole \; resorcinol}}{110.11 \; \cancel{g \; resorcinol}} \times \frac{1 \; \cancel{mole \; rd}}{1 \; \cancel{mole \; resorcinol}}$$

$$\times \frac{194.19 \; g \; rd}{1 \; \cancel{mole \; rd}} \approx 6.61 \; g \; resorcinol \; diacetate$$

Knowing the theoretical yield, we can now calculate the percent yield.

$$Percent \; Yield = \frac{amount \; of \; product \; obtained}{theoretical \; yield} \times 100$$

$$= \frac{4.11 \; \cancel{g \; resorcinol \; diacetate}}{6.61 \; \cancel{g \; resorcinol \; diacetate}} \times 100 \approx 62.2\%$$

Chapter 7

Significant figures

When calculating theoretical and percent yield, you may wonder how many digits to report. That question is answered by an understanding of significant figures, or those digits with meaning. *When calculating theoretical yield, it is the measured quantity that determines the number of significant figures in the answer.* The reason for this is that it is in this measurement of quantity that we must determine how many digits are meaningful. The measured amount of reactant is typically given as a mass or as a volume. Mass is usually measured on an electronic balance nowadays. When using an electronic device, all digits on the display are considered significant. Volume, on the other hand, is most commonly measured using a graduated cylinder in organic chemistry labs—though you will often use volumetric glassware in analytical laboratories. When using a graduated device, the measurement is made to the smallest gradation, and then one additional decimal place is estimated to arrive at the proper number of significant figures. After these measurements are made, theoretical yield calculations only involve multiplication and division by conversion factors, such as molar mass, density, and mole ratio. Conversion factors are taken to be exact and do not, therefore, impact the number of significant figures in the answer.

Let's revisit an example from the previous chapter: the synthesis of methyl salicylate from salicylic acid and methanol (Figure 7.1).

We'll highlight the role of the original measured quantities on the number of significant figures in the answer through two scenarios.

Scenario 1: Salicylic acid (5.05 grams) and methanol (20.50 mL) are heated with catalytic acid. If 3.77 mL of methyl salicylate are isolated, what is the percent yield?

The chemical equation is balanced as shown, so we need not insert any stoichiometric coefficients. As is often the case, we'll need two calculations to determine limiting reactant and theoretical yield. The first calculation determines the amount of methyl salicylate (abbreviated as "ms") that can be prepared from 5.05 grams of salicylic acid (abbreviated as "sa"). The provided mass of salicylic acid has three significant figures. Everything

DOI: 10.1201/9781003587408-7

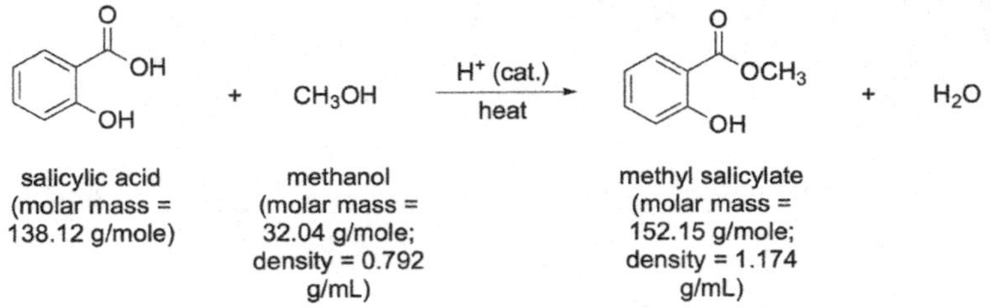

salicylic acid
(molar mass =
138.12 g/mole)

methanol
(molar mass =
32.04 g/mole;
density = 0.792
g/mL)

methyl salicylate
(molar mass =
152.15 g/mole;
density = 1.174
g/mL)

Figure 7.1 Methyl salicylate synthesis.

after that in the calculation is a conversion factor, which is taken to be exact and does not impact the number of significant figures in the answer. Consequently, the answer is rounded to three significant figures.

$$5.05 \; \cancel{g \; sa} \times \frac{1 \; \cancel{mol \; sa}}{138.12 \; \cancel{g \; sa}} \times \frac{1 \; \cancel{mol \; ms}}{1 \; \cancel{mol \; sa}} \times \frac{152.15 \; \cancel{g \; ms}}{1 \; \cancel{mol \; ms}}$$

$$\times \frac{1 \; mL \; ms}{1.174 \; \cancel{g \; ms}} \approx 4.74 \; mL \; methyl \; salicylate$$

The second calculation ascertains the amount of methyl salicylate that can be prepared using 20.50 mL of methanol (abbreviated as "m"). Since the measured volume of methanol contains four significant figures, the answer is rounded to four significant figures as well.

$$20.50 \; \cancel{mL \; m} \times \frac{0.792 \; \cancel{g \; m}}{1 \; \cancel{mL \; m}} \times \frac{1 \; \cancel{mol \; m}}{32.04 \; \cancel{g \; m}} \times \frac{1 \; \cancel{mol \; ms}}{1 \; \cancel{mol \; m}} \times \frac{152.15 \; \cancel{g \; ms}}{1 \; \cancel{mol \; ms}}$$

$$\times \frac{1 \; mL \; ms}{1.174 \; \cancel{g \; ms}} \approx 65.67 \; mL \; methyl \; salicylate$$

Our calculations show that salicylic acid is the limiting reactant and the theoretical yield is 4.74 mL of methyl salicylate. We can now determine the percent yield. This calculation contains two measured values. In multiplication and division, the value with fewer significant figures determines the number of significant digits in the answer. In this case, both the numerator and the denominator have three significant figures, so the answer should as well.

$$Percent \; Yield = \frac{amount \; of \; product \; obtained}{theoretical \; yield} \times 100$$

$$= \frac{3.77 \; \cancel{mL \; methyl \; salicylate}}{4.74 \; \cancel{mL \; methyl \; salicylate}} \times 100 \approx 79.5\%$$

Scenario 2: Salicylic acid (10.1735 grams) and methanol (3.00 mL) are heated with catalytic acid. If 8.72 mL of methyl salicylate are isolated, what is the percent yield?

In the calculation of the volume of methyl salicylate that could be made from 10.1735 grams salicylic acid, there are six significant figures in the given mass, so the answer should have six significant figures as well.

$$10.1735 \; \cancel{g \; sa} \times \frac{1 \; \cancel{mol \; sa}}{138.12 \; \cancel{g \; sa}} \times \frac{1 \; \cancel{mol \; ms}}{1 \; \cancel{mol \; sa}} \times \frac{152.15 \; \cancel{g \; ms}}{1 \; \cancel{mol \; ms}} \times \frac{1 \; mL \; ms}{1.174 \; \cancel{g \; ms}}$$
$$\approx 9.54592 \; mL \; methyl \; salicylate$$

In the calculation of the volume of methyl salicylate that could be generated from 3.00 mL of methanol, the given volume of methanol has only three significant digits, so the answer has only three as well.

$$3.00 \; \cancel{mL \; m} \times \frac{0.792 \; \cancel{g \; m}}{1 \; \cancel{mL \; m}} \times \frac{1 \; \cancel{mol \; m}}{32.04 \; \cancel{g \; m}} \times \frac{1 \; \cancel{mol \; ms}}{1 \; \cancel{mol \; m}} \times \frac{152.15 \; \cancel{g \; ms}}{1 \; \cancel{mol \; ms}}$$
$$\times \frac{1 mL \; ms}{1.174 \; \cancel{g \; ms}} \approx 9.61 \; mL \; methyl \; salicylate$$

The limiting reactant is salicylic acid once again, but the theoretical yield (9.54592 mL) now has more significant figures. As we calculate percent yield, remember that, when multiplying and dividing, the value with fewer significant digits dictates the proper number of significant figures in the answer. Since the numerator has only three significant digits, so should our answer.

$$Percent \; Yield = \frac{amount \; of \; product \; obtained}{theoretical \; yield} \times 100$$
$$= \frac{8.72 \; \cancel{mL \; methyl \; salicylate}}{9.54592 \; \cancel{mL \; methyl \; salicylate}} \times 100 \approx 91.3\%$$

CHAPTER 7 PRACTICE PROBLEMS

1. The Diels–Alder reaction is a classic cycloaddition between a diene (cyclopentadiene, in this case) and a dienophile (methyl acrylate, in the reaction shown in Figure 7.2). The product of this reaction is simply referred to as a Diels–Alder adduct. If 5.1 mL of cyclopentadiene and 6.25 mL of methyl acrylate are combined to yield 5.1357 grams of the Diels–Alder adduct, what is the percent yield for this reaction?

2. Consider the $S_N 2$ reaction between benzyl bromide and sodium cyanide to produce benzyl cyanide (Figure 7.3). If 5.65 mL of benzyl

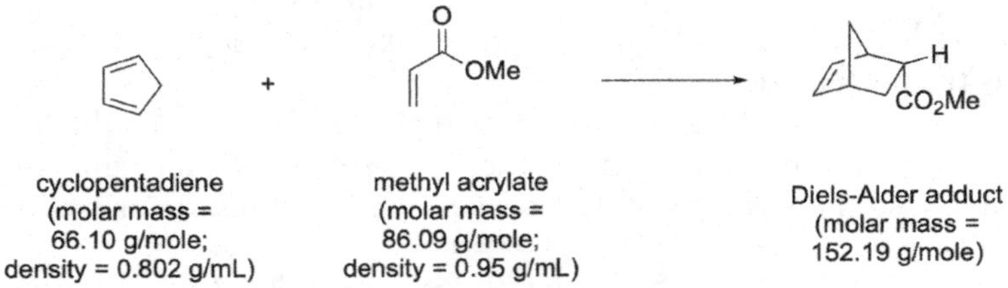

Figure 7.2 Diels–Alder reaction between cyclopentadiene and methyl acrylate.

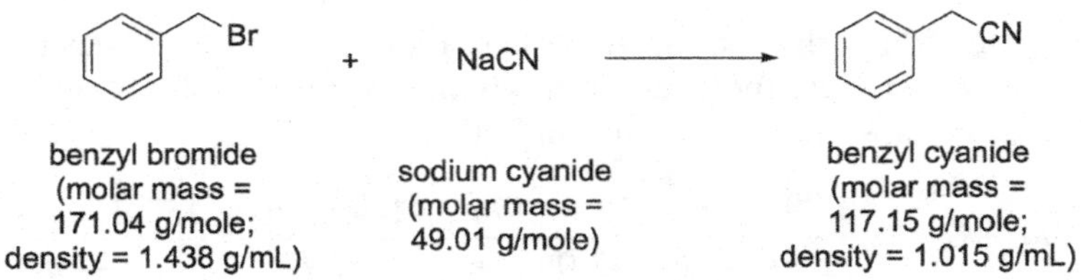

cyclopentadiene
(molar mass =
66.10 g/mole;
density = 0.802 g/mL)

methyl acrylate
(molar mass =
86.09 g/mole;
density = 0.95 g/mL)

Diels-Alder adduct
(molar mass =
152.19 g/mole)

benzyl bromide
(molar mass =
171.04 g/mole;
density = 1.438 g/mL)

sodium cyanide
(molar mass =
49.01 g/mole)

benzyl cyanide
(molar mass =
117.15 g/mole;
density = 1.015 g/mL)

Figure 7.3 S_N2 reaction between benzyl bromide and sodium cyanide.

bromide and 2.500 grams of sodium cyanide yield 4.8 mL of benzyl cyanide, what is the percent yield?

SOLUTIONS TO CHAPTER 7 PRACTICE PROBLEMS

1. This chemical equation is balanced as given, so we can jump directly into the calculations. The given volume of cyclopentadiene (abbreviated as "cpd") has only two significant figures, so the amount of Diels–Alder adduct (abbreviated as "DAa") that could be prepared from it contains only two significant figures as well.

$$5.1 \; mL \; cpd \times \frac{0.802 \; g \; cpd}{1 \; mL \; cpd} \times \frac{1 \; mole \; cpd}{66.10 \; g \; cpd} \times \frac{1 \; mole \; DAa}{1 \; mole \; cpd} \times \frac{152.19 \; g \; DAa}{1 \; mole \; DAa}$$
$$= 9.4 \; g \; Diels - Alder \; adduct$$

However, the given volume of methyl acrylate (abbreviated as "ma") has three significant figures, so the second calculation's answer has three significant digits.

$$6.25 \; mL \; ma \times \frac{0.95 \; g \; ma}{1 \; mL \; ma} \times \frac{1 \; mole \; ma}{86.09 \; g \; ma} \times \frac{1 \; mole \; DAa}{1 \; mole \; ma}$$
$$\times \frac{152.19 \; g \; DAa}{1 \; mole \; DAa} = 10.5 \; g \; Diels - Alder \; adduct$$

The limiting reactant for this transformation is cyclopentadiene, and the theoretical yield is 9.4 grams of Diels–Alder adduct. When calculating the percent yield, the two significant figures in the denominator limit the number of significant digits in the answer to two.

$$Percent\ Yield = \frac{amount\ of\ product\ obtained}{theoretical\ yield} \times 100$$

$$= \frac{5.1357\ \cancel{g\ Diels-Alder\ adduct}}{9.4\ \cancel{g\ Diels-Alder\ adduct}} \times 100 \approx 55\%$$

2. This chemical equation was not balanced as given, but the unstated inorganic by-product that balances the reaction (sodium bromide) has no impact on our calculations (Figure 7.4).

There are three significant figures in the given volume of benzyl bromide (abbreviated as "bb"), so the answer to the first calculation will have three significant digits. Note that benzyl cyanide is abbreviated as "bc."

$$5.65\ \cancel{mL\ bb} \times \frac{1.438\ \cancel{g\ bb}}{1\ \cancel{mL\ bb}} \times \frac{1\ \cancel{mole\ bb}}{171.04\ \cancel{g\ bb}} \times \frac{1\ \cancel{mole\ bc}}{1\ \cancel{mole\ bb}} \times \frac{117.15\ \cancel{g\ bc}}{1\ \cancel{mole\ bc}}$$

$$\times \frac{1\ mL\ bc}{1.015\ \cancel{g\ bc}} = 5.48\ mL\ benzyl\ cyanide$$

However, there are four significant figures in the given mass of sodium cyanide (NaCN), so the outcome of the second calculation contains four significant digits.

$$2.500\ \cancel{g\ NaCN} \times \frac{1\ \cancel{mole\ NaCN}}{49.01\ \cancel{g\ NaCN}} \times \frac{1\ \cancel{mole\ bc}}{1\ \cancel{mole\ NaCN}} \times \frac{117.15\ \cancel{g\ bc}}{1\ \cancel{mole\ bc}}$$

$$\times \frac{1\ mL\ bc}{1.015\ \cancel{g\ bc}} = 5.888\ mL\ benzyl\ cyanide$$

The limiting reactant is benzyl bromide, and the theoretical yield is 5.48 mL of benzyl cyanide. When calculating the percent yield, there

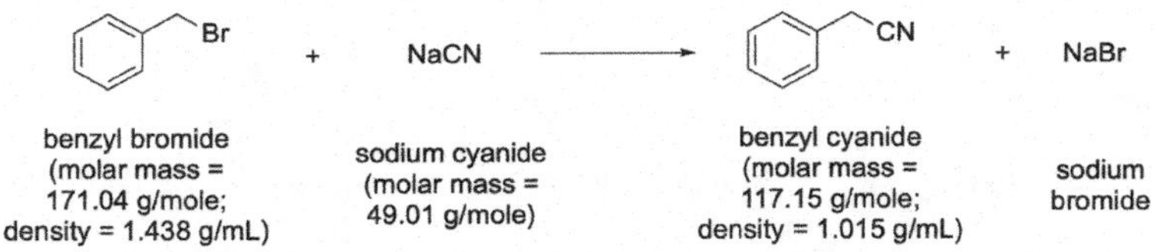

benzyl bromide
(molar mass =
171.04 g/mole;
density = 1.438 g/mL)

sodium cyanide
(molar mass =
49.01 g/mole)

benzyl cyanide
(molar mass =
117.15 g/mole;
density = 1.015 g/mL)

sodium
bromide

Figure 7.4 Balanced S_N2 reaction between benzyl bromide and sodium cyanide.

are only two significant figures in the numerator, which limits the number of significant digits in the answer to two.

$$Percent\ Yield = \frac{amount\ of\ product\ obtained}{theoretical\ yield} \times 100$$

$$= \frac{4.8\ \cancel{mL\ benzyl\ cyanide}}{5.48\ \cancel{mL\ benzyl\ cyanide}} \times 100 \approx 88\%$$

Planning a synthesis

How much reactant to use?

It is undoubtedly useful to be able to calculate theoretical yield, but there is another—arguably even more useful—skill that is closely related to it. That is calculating the amount of reactant(s) to use in order to generate a desired amount of product. You could think of this as the reverse of a theoretical yield calculation.

In a research setting, it would seem unlikely that you'd be mixing together particular, prescribed amounts of reactants. After all, if you are conducting a brand new reaction that no one has ever done before, who would tell you how much of each reactant to use? You'd have to decide that for yourself, and the way you'd do that would be based on how much of the target product you hope to produce.

For a case study, let's revisit the very first chemical reaction we considered in Chapter 3: the synthesis of aspirin (Figure 8.1).

Imagine that your goal is to produce 5.00 grams of aspirin. How much salicylic acid and acetic anhydride would you need in order to achieve that goal? To answer this question, we'll need to perform two calculations, each beginning with the target quantity of aspirin. From our prior experience, we already know that this equation is balanced as written. In each case, we convert the mass of aspirin desired to moles of aspirin. Then, we use the appropriate mole ratio to convert to moles of a reactant. Finally, we convert moles of that reactant into grams if it is a solid or milliliters if it is a liquid. In the calculations that follow, salicylic acid is abbreviated as "sa" and acetic anhydride is abbreviated as "aa."

$$5.00 \; \cancel{g \; aspirin} \times \frac{1 \; \cancel{mole \; aspirin}}{180.16 \; \cancel{g \; aspirin}} \times \frac{1 \; \cancel{mole \; sa}}{1 \; \cancel{mole \; aspirin}}$$

$$\times \frac{138.12 \; g \; sa}{1 \; \cancel{mole \; sa}} \approx 3.83 \; g \; salicylic \; acid$$

DOI: 10.1201/9781003587408-8

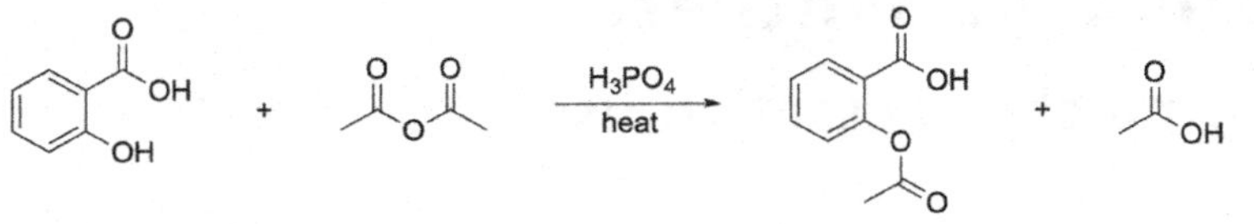

salicylic acid
(molar mass =
138.12 g/mole)

acetic anhydride
(molar mass =
102.09 g/mole;
density = 1.082 g/mL)

aspirin
(molar mass =
180.16 g/mole)

acetic acid

Figure 8.1 Aspirin synthesis.

$$5.00 \ \cancel{g \ aspirin} \times \frac{1 \ \cancel{mole \ aspirin}}{180.16 \ \cancel{g \ aspirin}} \times \frac{1 \ \cancel{mole \ aa}}{1 \ \cancel{mole \ aspirin}} \times \frac{102.09 \ \cancel{g \ aa}}{1 \ \cancel{mole \ aa}}$$

$$\times \frac{1 \ mL \ aa}{1.082 \ \cancel{g \ aa}} \approx 2.62 \ mL \ acetic \ anhydride$$

So, we would need 3.83 grams of salicylic acid and 2.62 mL of acetic anhydride in order to produce 5.00 grams of aspirin. Sometimes organic chemists use a slight excess of the cheaper reactant to ensure the complete conversion of the more valuable reactant into product. It will be easier to ascertain which reactant is more precious as your experience in the field grows. However, for the time being, you can often surmise that the larger, more complex structure is the more valuable reactant. Therefore, it would be an educated guess that, in this instance, salicylic acid is the more precious commodity, and we might use a slight excess (perhaps 1.1 times the needed amount) of acetic anhydride to ensure its complete conversion to product. Therefore, we might actually use 3.83 grams of salicylic acid and 2.88 mL of acetic anhydride for the synthesis.

Of course, the percent yield is not typically a perfect 100%, so we may fall a bit short of our goal. Let's say that our aspirin synthesis produces 4.78 grams. Then, the percent yield for the synthesis is 95.6%. Remember, you don't need to calculate the theoretical yield in this case because you planned your synthesis so that the theoretical yield would be 5.00 grams of aspirin.

$$Percent \ Yield = \frac{amount \ of \ product \ obtained}{theoretical \ yield} \times 100$$

$$= \frac{4.78 \ \cancel{g \ aspirin}}{5.00 \ \cancel{g \ aspirin}} \times 100 = 95.6\%$$

Now, we are ready to use the principles of theoretical yield calculation to plan our own organic syntheses!

CHAPTER 8 PRACTICE PROBLEMS

1. Consider the Fischer esterification of phenylacetic acid with methanol using catalytic acid and heat to produce methyl phenylacetate, an ester with a honey-like aroma (Figure 8.2).
   (a) How much of each reactant would you need to use if you wanted to generate 10.00 mL of methyl phenylacetate?
   (b) The Fischer esterification is a freely reversible reaction. For this reason, we must drive the reaction to completion using an excess of one reactant. This is an example of using Le Châtelier's principle to push the equilibrium to the right. Since methanol is the cheaper reactant, we'll use it in excess. Adjust your calculation from part (a) to reflect the use of a sixfold excess of methanol.
   (c) Imagine that you've performed this reaction several times before, and as a result of your prior experience, you know that it tends to give about an 80% yield. How might you scale up the amount of each reactant (while maintaining the sixfold excess of methanol) so that you are more likely to actually obtain 10.00 mL of methyl phenylacetate?

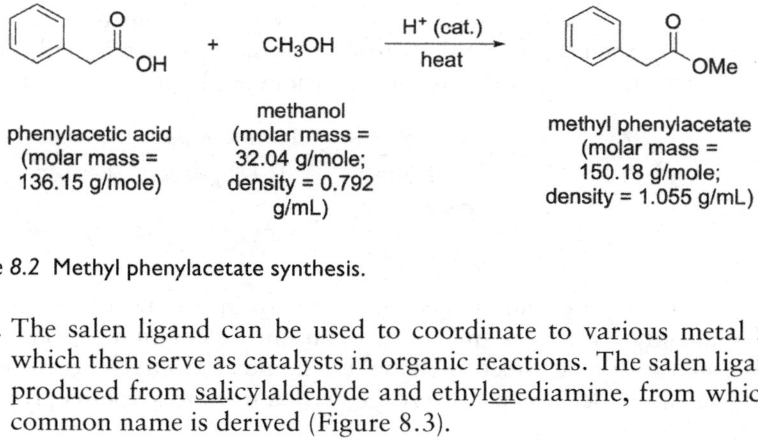
O
OH
+
CH3OH
H+ (cat.)
heat
O
OMe
phenylacetic acid
(molar mass =
136.15 g/mole)
methanol
(molar mass =
32.04 g/mole;
density = 0.792
g/mL)
methyl phenylacetate
(molar mass =
150.18 g/mole;
density = 1.055 g/mL)

Figure 8.2 Methyl phenylacetate synthesis.

2. The salen ligand can be used to coordinate to various metal ions, which then serve as catalysts in organic reactions. The salen ligand is produced from salicylaldehyde and ethylenediamine, from which its common name is derived (Figure 8.3).

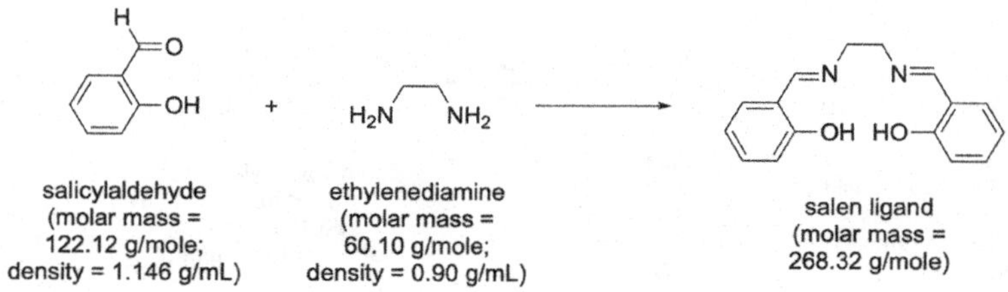
H
O
OH
+
H2N
NH2
N
N
OH HO
salicylaldehyde
(molar mass =
122.12 g/mole;
density = 1.146 g/mL)
ethylenediamine
(molar mass =
60.10 g/mole;
density = 0.90 g/mL)
salen ligand
(molar mass =
268.32 g/mole)

Figure 8.3 Salen ligand synthesis.

How much salicylaldehyde and ethylenediamine do you need to use if the goal is to prepare 5.00 grams of salen ligand? If you've performed the reaction before and obtained about 90% yield, how can you adjust for that to ensure that you obtain the desired amount of salen?

SOLUTIONS TO CHAPTER 8 PRACTICE PROBLEMS

1. (a) The equation we were given was not balanced. However, when we do balance the equation, we find that the water by-product was merely omitted, and no stoichiometric coefficients are necessary (Figure 8.4). So, in this instance, balancing the equation will have no impact on our calculations.

   The first calculation will determine the amount of phenylacetic acid (abbreviated "paa") needed. This calculation begins with the given target volume of methyl phenylacetate (abbreviated "mpa"). Since we were given a volume of product (rather than a mass), the first step is converting volume to mass using density. Then, the desired mass of methyl phenylacetate is converted to moles using molar mass. The mole ratio allows us to convert the desired moles of methyl phenylacetate to the needed moles of phenylacetic acid. Lastly, the necessary number of moles of phenylacetic acid is converted to the needed mass of this reactant, about 9.564 grams.

$$10.00 \; \cancel{mL \; mpa} \times \frac{1.055 \; \cancel{g \; mpa}}{1 \; \cancel{mL \; mpa}} \times \frac{1 \; \cancel{mole \; mpa}}{150.18 \; \cancel{g \; mpa}} \times \frac{1 \; \cancel{mole \; paa}}{1 \; \cancel{mole \; mpa}}$$

$$\times \frac{136.15 \; g \; paa}{1 \; \cancel{mole \; paa}} \approx 9.564 \; g \; phenylacetic \; acid$$

A similar calculation is used to determine the amount of methanol (abbreviated as "m") needed, but this calculation has one extra step at the end. Since methanol is a liquid at room temperature, we must use density to convert the needed mass to a volume, roughly 2.842 mL.

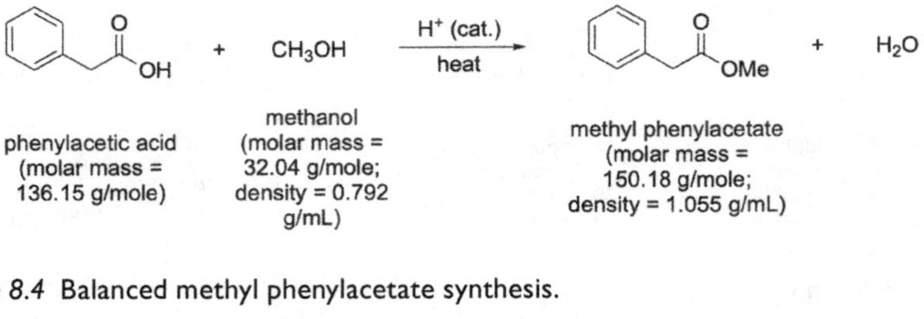

| phenylacetic acid (molar mass = 136.15 g/mole) | + | methanol (molar mass = 32.04 g/mole; density = 0.792 g/mL) | $\xrightarrow[\text{heat}]{\text{H}^+ \text{ (cat.)}}$ | methyl phenylacetate (molar mass = 150.18 g/mole; density = 1.055 g/mL) | + | H_2O |

Figure 8.4 Balanced methyl phenylacetate synthesis.

$$10.00 \; \cancel{mL \; mpa} \times \frac{1.055 \; \cancel{g \; mpa}}{1 \; \cancel{mL \; mpa}} \times \frac{1 \; \cancel{mole \; mpa}}{150.18 \; \cancel{g \; mpa}} \times \frac{1 \; \cancel{mole \; m}}{1 \; \cancel{mole \; mpa}}$$

$$\times \frac{32.04 \; \cancel{g \; m}}{1 \; \cancel{mole \; m}} \times \frac{1 \; mL \; m}{0.792 \; \cancel{g \; m}} \approx 2.842 \; mL \; methanol$$

(b) Using a six-fold excess of methanol simply means that we'll utilize six times more methanol than we need for this reaction.

$$2.842 \; mL \; methanol \times 6 \approx 17.05 \; mL \; methanol$$

The amount of phenylacetic acid to be used remains unchanged: 9.564 grams.

(c) If you know that you tend to obtain only about 80% yield, you can simply divide the needed amount of each reactant by 0.80 to scale up accordingly.

$$\frac{9.564 \; g \; phenylacetic \; acid}{0.80} \approx 11.96 \; g \; phenylacetic \; acid$$

$$\frac{17.05 \; mL \; methanol}{0.80} \approx 21.31 \; mL \; methanol$$

Using these amounts of the reactants makes it more likely that you will achieve your goal of 10.00 mL of methyl phenylacetate. If that isn't completely clear to you, let's prove it to ourselves. We already know that methanol is used in a sixfold excess, so phenylacetic acid is the limiting reactant. Let's calculate how much methyl phenylacetate we could make from the 11.96 grams of phenylacetic acid we've decided to use.

$$11.96 \; \cancel{g \; paa} \times \frac{1 \; \cancel{mole \; paa}}{136.15 \; \cancel{g \; paa}} \times \frac{1 \; \cancel{mole \; mpa}}{1 \; \cancel{mole \; paa}} \times \frac{150.18 \; \cancel{g \; mpa}}{1 \; \cancel{mole \; mpa}}$$

$$\times \frac{1 \; mL \; mpa}{1.055 \; \cancel{g \; mpa}} \approx 12.50 \; mL \; methyl \; phenylacetate$$

That theoretical yield is more than the 10.00 mL we wanted, but don't forget that our prior experience shows that the reaction usually only gives about 80% yield. Let's fill what we know into the equation for percent yield.

$$Percent \; Yield = \frac{amount \; of \; product \; obtained}{theoretical \; yield} \times 100$$

$$80\% = \frac{amount \; of \; methyl \; phenylacetate \; obtained}{12.50 \; mL \; methyl \; phenylacetate} \times 100$$

Now, let's solve for the unknown, which is the amount of methyl phenylacetate we'll actually obtain.

$$\frac{80\% \times \left(12.50 \; mL \; methyl \; phenylacetate\right)}{100}$$

$$= amount \; of \; methyl \; phenylacetate \; obtained$$

$$10.00 \; mL \; methyl \; phenylacetate =$$
$$amount \; of \; methyl \; phenylacetate \; obtained$$

Now we've proven that, after accounting for the anticipated percent yield, we can expect to actually obtain 10.00 mL of methyl phenylacetate.

2. Did you perform your calculations without first balancing the chemical equation? If so, go back and balance the equation. Then, adjust your calculations accordingly before proceeding.

Since the chemical equation was not balanced as given, we must first remedy that situation before proceeding to the calculations. By inspection of the structures, it quickly becomes clear that we must place a stoichiometric coefficient of 2 in front of salicylaldehyde. Then, we'll observe that two molecules of water are produced as a by-product of this reaction (Figure 8.5).

Now we are able to proceed to the calculations. Let's begin with the calculation of the needed amount of salicylaldehyde ("sal"). We start by converting the desired mass of salen into moles of salen using the molar mass. Then, it is critical to include the correct mole ratio as we convert moles of salen into moles of salicylaldehyde (1:2). The number of moles of salicylaldehyde that we need is then converted to a mass using the molar mass and finally to a volume using the density.

$$5.00 \; \cancel{g \; salen} \times \frac{1 \; \cancel{mole \; salen}}{268.32 \; \cancel{g \; salen}} \times \frac{2 \; \cancel{mole \; sal}}{1 \; \cancel{mole \; salen}} \times \frac{122.12 \; \cancel{g \; sal}}{1 \; \cancel{mole \; sal}}$$

$$\times \frac{1 \; mL \; sal}{1.146 \; \cancel{g \; sal}} \approx 3.97 \; mL \; salicylaldehyde$$

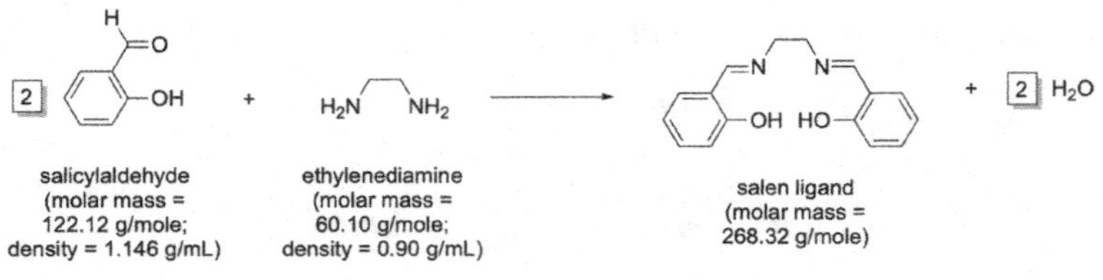

Figure 8.5 Balanced salen ligand synthesis.

The calculation of the necessary volume of ethylenediamine ("en") closely mirrors the calculation above, except, of course, that the mole ratio, molar mass of ethylenediamine, and density have different values.

$$5.00 \; \cancel{g \; salen} \times \frac{1 \; \cancel{mole \; salen}}{268.32 \; \cancel{g \; salen}} \times \frac{1 \; \cancel{mole \; en}}{1 \; \cancel{mole \; salen}}$$

$$\times \frac{60.10 \; \cancel{g \; en}}{1 \; \cancel{mole \; en}} \times \frac{1 \; mL \; en}{0.90 \; \cancel{g \; en}} \approx 1.24 \; mL \; ethylenediamine$$

If we typically obtain a 90% yield for this reaction, we can simply divide each quantity of reactant by 0.90 to adjust the values so that we are more likely to achieve our goal of obtaining 5.00 grams of salen.

$$\frac{3.97 \; mL \; salicylaldehyde}{0.90} \approx 4.41 \; mL \; salicylaldehyde$$

$$\frac{1.24 \; mL \; ethylenediamine}{0.90} \approx 1.38 \; mL \; ethylenediamine$$

So, we'll begin our synthesis with 4.41 mL of salicylaldehyde and 1.38 mL of ethylenediamine, and we should obtain pretty close to 5.00 grams of salen if the percent yield is typically close to 90%. If you still aren't quite convinced, we can prove it to ourselves using theoretical yield calculations. Since we have perfectly matched quantities of reactants in this case, both calculations yield about the same theoretical yield: 5.55 grams of salen.

$$4.41 \; \cancel{mL \; sal} \times \frac{1.146 \; \cancel{g \; sal}}{1 \; \cancel{mL \; sal}} \times \frac{1 \; \cancel{mole \; sal}}{122.12 \; \cancel{g \; sal}} \times \frac{1 \; \cancel{mole \; salen}}{2 \; \cancel{moles \; sal}} \times \frac{268.32 \; g \; salen}{1 \; \cancel{mole \; salen}}$$

$$\approx 5.55 \; g \; salen$$

$$1.38 \; \cancel{mL \; en} \times \frac{0.90 \; \cancel{g \; en}}{1 \; \cancel{mL \; en}} \times \frac{1 \; \cancel{mole \; en}}{60.10 \; \cancel{g \; en}} \times \frac{1 \; \cancel{mole \; salen}}{1 \; \cancel{moles \; en}} \times \frac{268.32 \; g \; salen}{1 \; \cancel{mole \; salen}}$$

$$\approx 5.55 \; g \; salen$$

That's more than we set out to prepare, but remember that this reaction usually gives us a 90% yield. Once we account for this, we can see that we are indeed likely to end up with roughly 5.00 grams of salen in hand.

$$Percent \; Yield = \frac{amount \; of \; product \; obtained}{theoretical \; yield} \times 100$$

$$90\% = \frac{amount \; of \; product \; obtained}{5.55 \; g \; salen} \times 100$$

$$\frac{90\% \times (5.55 \ g \ salen)}{100} = amount \ of \ product \ obtained$$

$$5.00 \ g \ salen \approx amount \ of \ product \ obtained$$

Chapter 9

Practice problems

1. In the S_N2 reaction shown in Figure 9.1, an alkyl iodide reacts with a thiolate (i.e., the conjugate base of a thiol) to yield a thioether (an ether with sulfur instead of oxygen linking the two alkyl groups). What is the percent yield for the reaction if we begin with 8.25 mL of cyclohexyl iodide and 4.50 grams of sodium methanethiolate and obtain 6.99 grams of cyclohexyl methyl sulfide?

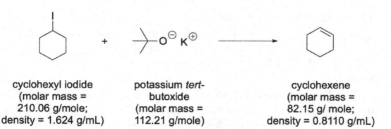

cyclohexyl iodide
(molar mass =
210.06 g/mole;
density = 1.624 g/mL)

sodium
methanethiolate
(molar mass =
70.08 g/mole)

cyclohexyl
methyl sulfide
(molar mass =
130.25 g/ mole)

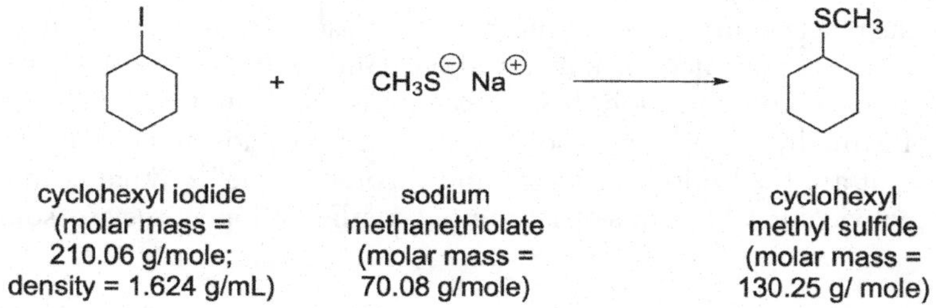

Figure 9.1 S_N2 reaction between cyclohexyl iodide and sodium methanethiolate.

2. The same reactant used in Problem 1, cyclohexyl iodide, can be a substrate for an E2 reaction instead (Figure 9.2). If it is treated with potassium *tert*-butoxide, the product will be cyclohexene. Such a reaction uses 8.25 mL of cyclohexyl iodide and 7.25 grams of potassium *tert*-butoxide, yielding 6.12 mL of cyclohexene. What is the percent yield?

cyclohexyl iodide
(molar mass =
210.06 g/mole;
density = 1.624 g/mL)

potassium *tert*-
butoxide
(molar mass =
112.21 g/mole)

cyclohexene
(molar mass =
82.15 g/ mole;
density = 0.8110 g/mL)

Figure 9.2 E2 reaction between cyclohexyl iodide and potassium *tert*-butoxide.

DOI: 10.1201/9781003587408-9

3. The dehydration of 2-pentanol can be accomplished by heating with catalytic sulfuric acid (Figure 9.3). This reaction produces a mixture of three products. The major product is *trans*-2-pentene, and there are two minor products: *cis*-2-pentene and 1-pentene. If we dehydrate 40.00 mL of 2-pentanol and obtain 15.2 grams of the pentenes product mixture, what is the percent yield? (*Hint:* One molecule of 2-pentanol can yield only one molecule of alkene, regardless of which one it is.)

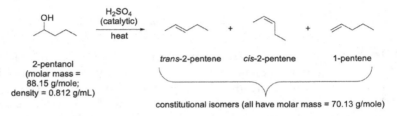

Figure 9.3 Dehydration of 2-pentanol.

4. The reaction in Figure 9.4 illustrates the use of phosphorus tribromide to convert an alcohol into an alkyl halide with inversion of stereo-chemical configuration. If the reaction is conducted with 2.25 grams of 2-methyl-1-cyclopentanol and 2.25 mL of phosphorus tribromide, what are the limiting reactant and theoretical yield? What is the percent yield if 3.11 grams of 1-bromo-2-methylcyclopentane are isolated?

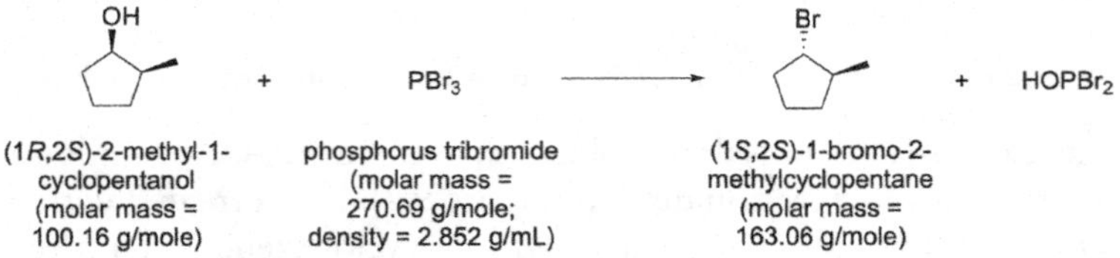

Figure 9.4 Reaction of an alcohol with PBr_3.

5. The Williamson ether synthesis is a reaction of alcohols. The alcohol is first treated with sodium hydride to generate its conjugate base, which then acts as a nucleophile in a subsequent S_N2 reaction with an unhindered alkyl halide. The complete, balanced equation is given in Figure 9.5.

The Williamson ether synthesis shown in Figure 9.5 is performed with 5.00 mL of (*S*)-1-phenylethanol, 1.000 gram of sodium hydride, and 3.70 mL of propyl chloride. What are the limiting reactant and theoretical yield? What is the percent yield if the amount of (S)-(1-propoxyethyl)benzene generated is 6.251 grams?

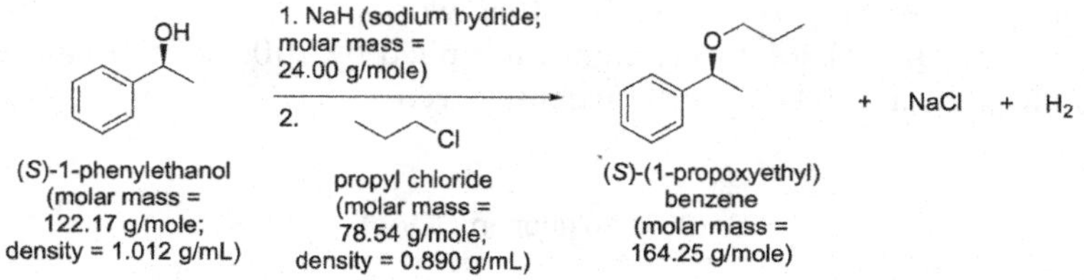

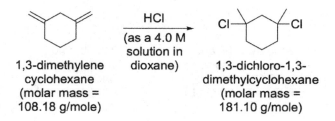

(S)-1-phenylethanol
(molar mass =
122.17 g/mole;
density = 1.012 g/mL)

1. NaH (sodium hydride;
molar mass =
24.00 g/mole)

2.

propyl chloride
(molar mass =
78.54 g/mole;
density = 0.890 g/mL)

(S)-(1-propoxyethyl)
benzene
(molar mass =
164.25 g/mole)

+   NaCl   +   H₂

Figure 9.5 A Williamson ether synthesis.

6. The opening of styrene oxide, an epoxide, under basic conditions with dimethylamine is shown in Figure 9.6. Consider an experiment using 6.75 mL of (S)-styrene oxide and 4.00 mL of dimethylamine. What is the percent yield if this experiment produces 8.76 grams of (S)-2-(dimethylamino)-1-phenylethanol?

(S)-styrene oxide
(molar mass =
120.15 g/mole;
density = 1.051 g/mL)

+

H₃C   N   CH₃
         H
dimethylamine
(molar mass =
45.09 g/mole;
density = 0.680 g/mL)

(S)-2-(dimethylamino)-1-
phenylethanol
(molar mass =
165.24 g/mole)

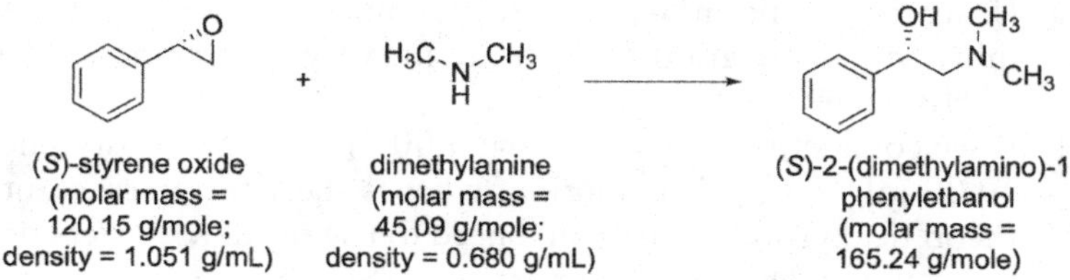

Figure 9.6 Basic epoxide opening.

7. Ionic hydrohalogenation of alkenes is a reaction in which HX (where X is a halogen) adds across the carbon–carbon pi bond. Consider the example of ionic hydrohalogenation shown in Figure 9.7. Imagine beginning with 11.50 grams of 1,3-dimethylenecyclohexane and 54.0 mL of a 4.0 M solution of hydrochloric acid in a solvent called dioxane. What is the percent yield for this transformation if 15.3 grams of 1,3-dichloro-1,3-dimethylcyclohexane are generated?

1,3-dimethylene
cyclohexane
(molar mass =
108.18 g/mole)

HCl

(as a 4.0 M
solution in
dioxane)

1,3-dichloro-1,3-
dimethylcyclohexane
(molar mass =
181.10 g/mole)

Figure 9.7 An ionic hydrohalogenation.

8. One way to achieve the anti-Markovnikov hydrobromination of 2-bromostyrene involves treatment with a 33% w/w solution of HBr in acetic acid (HOAc), as shown in Figure 9.8. The designation "w/w" means weight percentage. In other words, it is the ratio of solute weight to total solution weight multiplied by 100. (Note that weight and mass are used interchangeably here.)

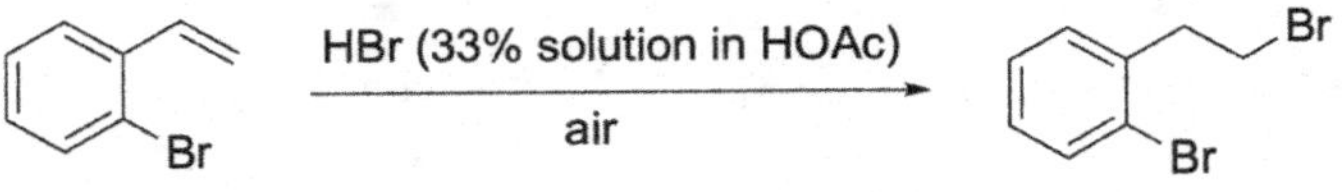

| 2-bromostyrene (molar mass = 183.05 g/mole; density = 1.460 g/mL) | (molar mass of HBr = 80.91 g/mole; density of solution = 1.40 g/mL) | 1-bromo-2-(2-bromoethyl)benzene (molar mass = 263.96 g/mole) |

Figure 9.8 Anti-Markovnikov hydrobromination of 2-bromostyrene.

(a) Using the definition of a 33% w/w solution provided above, determine how many grams of HBr are in every 1,000 grams (or 1 kg) of HBr solution.
(b) If we conduct the reaction using 5.00 mL of 2-bromostyrene and 10.00 mL of the HBr solution, what is the limiting reactant and theoretical yield? (*Hint:* You'll need to use the new conversion factor that you determined in part (a).)
(c) What is the percent yield if 8.67 grams of product are obtained?
9. The halogenation of an alkene yields a vicinal (neighboring) dihalide. An example is shown in Figure 9.9. What is the percent yield for the reaction if 20.00 mL of 2-methyl-2-butene and 10.00 mL of Br_2 are used to prepare 35.31 grams of 2,3-dibromo-2-methylbutane? Note that molecular bromine (Br_2) has a molar mass of 159.81 g/mole and a density of 3.12 g/mL and dichloromethane (CH_2Cl_2) is an inert solvent for the reaction.

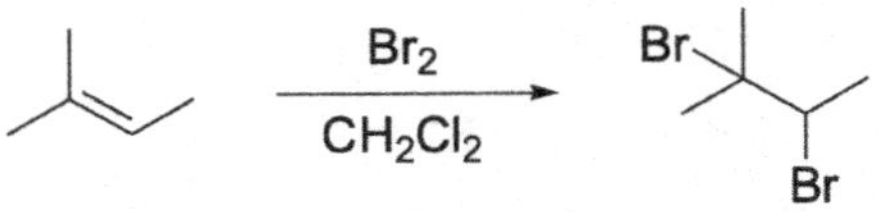

| 2-methyl-2-butene (molar mass = 70.13 g/mole; density = 0.662 g/mL) | 2,3-dibromo-2-methylbutane (molar mass = 229.94 g/mole) |

Figure 9.9 Halogenation of an alkene.

10. The epoxidation of styrene using the oxidant *m*CPBA (*meta*-chloroper-
    oxybenzoic acid) is shown in Figure 9.10. Note that the stoichiometry is
    correct as shown, although a by-product has been omitted for clarity.

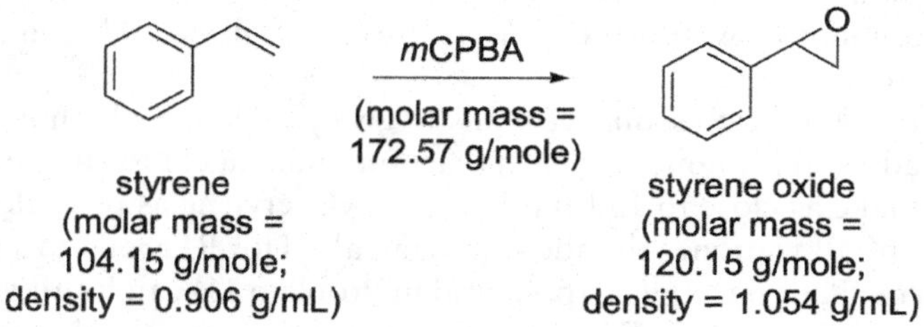

*m*CPBA

(molar mass =
172.57 g/mole)

styrene
(molar mass =
104.15 g/mole;
density = 0.906 g/mL)

styrene oxide
(molar mass =
120.15 g/mole;
density = 1.054 g/mL)

Figure 9.10 Epoxidation of styrene.

(a) If you wish to prepare 10.00 mL of styrene oxide, how much sty-
    rene and *m*CPBA must you use?
(b) The oxidant *m*CPBA is available as a solid that only contains
    77% of the desired reagent. How should you adjust the amount of
    *m*CPBA to account for this?
(c) If the reaction commonly gives you about 85% yield, how should
    you adjust the amounts of each reactant to ensure that you make
    as close to 10.00 mL of styrene oxide as possible?

11. Styrene can be converted into phenylacetylene through a three-step
    process (Figure 9.11). First, the addition of bromine across the car-
    bon–carbon pi bond yields a vicinal dibromide. Then, treatment with
    the strong base sodamide ($NaNH_2$) results in two E2 reactions to
    install the two pi bonds of the alkyne, followed by deprotonation of
    the resulting alkyne. Lastly, the addition of water reinstalls the proton
    on the terminal alkyne.

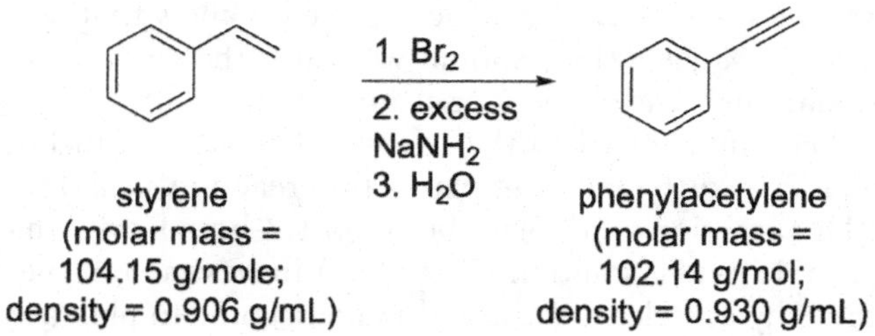

1. Br_2

2. excess
   $NaNH_2$
3. H_2O

styrene
(molar mass =
104.15 g/mole;
density = 0.906 g/mL)

phenylacetylene
(molar mass =
102.14 g/mol;
density = 0.930 g/mL)

Figure 9.11 Conversion of styrene into phenylacetylene.

(a) Plan the synthesis of 15.00 mL of phenylacetylene. Note that molecular bromine (Br_2) has a molar mass of 159.81 g/mole and a density of 3.12 g/mL and sodamide ($NaNH_2$) has a molar mass of 39.01 g/mole. Plan to use one molar equivalent of Br_2 and 3.1 molar equivalents of sodamide. You need not calculate a specific amount of water to use; a great excess will be added in the third step.

(b) If the reaction sequence tends to give you 90% yield, how can you adjust the amounts of each reactant and reagent to ensure that you make as close to 15.00 mL of phenylacetylene as possible.

12. Alkyne alkylation is the addition of an alkyl (or R) group to a terminal alkyne. Phenylacetylene (prepared in Problem 11) can be alkylated via a two-step sequence (Figure 9.12). The first step is the deprotonation of the alkyne using the potent base sodamide ($NaNH_2$). The second step is treatment with an unhindered electrophile, such as butyl bromide, that is suitable for S_N2 reaction.

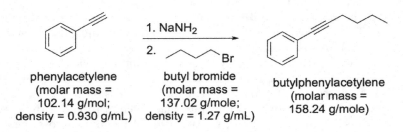

phenylacetylene
(molar mass =
102.14 g/mol;
density = 0.930 g/mL)

butyl bromide
(molar mass =
137.02 g/mole;
density = 1.27 g/mL)

butylphenylacetylene
(molar mass =
158.24 g/mole)

Figure 9.12 Alkylation of phenylacetylene.

(a) Plan the synthesis of 25.00 grams of butylphenylacetylene. Consider the stoichiometry to be correct as shown, although by-products have been omitted for clarity. And, note that sodamide ($NaNH_2$) has a molar mass of 39.01 g/mole.

(b) If by performing this sequence before we know that we can expect about 75% yield, how should we adjust the amounts of reactants to maximize the chances of attaining our goal?

13. We've been introduced to the Diels–Alder reaction in Problem 1 of Chapter 7. A different example of this reaction is shown in Figure 9.13. This reaction takes place between 1,3-butadiene, which is available as a 15 wt.% solution (15% w/w) in hexane (an inert solvent), and dimethylacetylene. We are ultimately going to plan a synthesis of 5.00 grams of the Diels–Alder adduct shown in Figure 9.13.

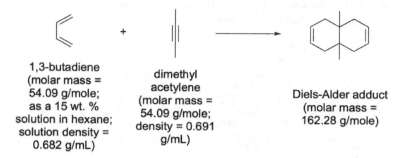

Figure 9.13 Diels-Alder reaction between 1,3-butadiene and dimethylacetylene.

(a) Balance the reaction equation. This will require careful examination of the structures. Labeling the carbon atoms using letters may help.

(b) Using our understanding of a weight percent solution developed in Problem 8, determine how many grams of 1,3-butadiene are in every 1,000 grams (or 1 kg) of 1,3-butadiene solution.

(c) Now, calculate the amount of each reactant necessary for the preparation of 5.00 grams of the Diels–Alder adduct.

14. Conjugated dienes, such as 1,3-butadiene, can undergo ionic hydrohalogenation so as to give a mixture of two products. The product that predominates depends on the temperature at which the reaction is conducted. If 6.75 mL of a 15% w/w solution of 1,3-butadiene and 2.25 mL of a 33% w/w solution of HBr undergo the reaction illustrated (Figure 9.14) to produce 1.53 grams of the product mixture, what is the percent yield?

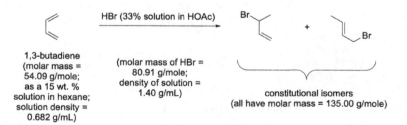

Figure 9.14 Ionic hydrohalogenation of 1,3-butadiene.

15. The Friedel–Crafts alkylation of toluene with ethyl chloride and catalytic aluminum trichloride is shown in Figure 9.15.

Plan a synthesis of approximately 5.00 grams of 2-ethyltoluene and 4-ethyltoluene using 0.1 molar equivalent of aluminum trichloride (a catalytic amount). Adjust for the fact that, in performing this reaction previously, you obtained about 70% yield.

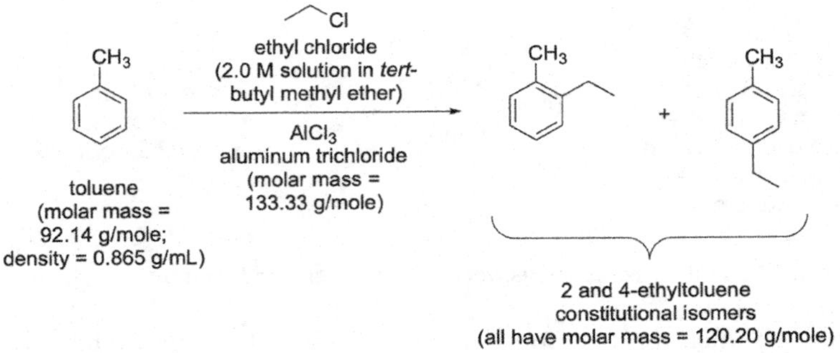

CH₃

Cl
ethyl chloride
(2.0 M solution in *tert*-butyl methyl ether)

AlCl₃
aluminum trichloride
(molar mass =
133.33 g/mole)

toluene
(molar mass =
92.14 g/mole;
density = 0.865 g/mL)

CH₃

+

CH₃

2 and 4-ethyltoluene
constitutional isomers
(all have molar mass = 120.20 g/mole)

Figure 9.15 Friedel–Crafts alkylation of toluene.

16. Treatment of cyclohexanone with dimethylamine and catalytic acid, while removing water, will yield the corresponding enamine (Figure 9.16). If 3.00 mL of cyclohexanone and 15.00 mL of a 2.0 M solution of dimethylamine in THF (tetrahydrofuran, an inert solvent) are used to produce 3.51 grams of the enamine, what is the percent yield for the reaction?

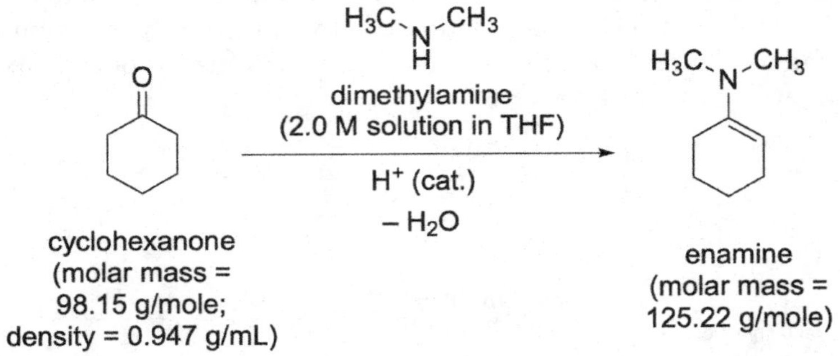

H_3C CH_3
N
H
dimethylamine
(2.0 M solution in THF)

H^+ (cat.)

$- H_2O$

O

cyclohexanone
(molar mass =
98.15 g/mole;
density = 0.947 g/mL)

H_3C CH_3
N

enamine
(molar mass =
125.22 g/mole)

Figure 9.16 Enamine formation from cyclohexanone.

17. The Baeyer–Villiger oxidation of 2,2-dimethylpropiophenone is shown in Figure 9.17. The stoichiometry of the reaction is correct as given, though an undesired by-product has been omitted for simplicity. Consider an experiment that uses 4.25 mL of 2,2-dimethylpropiophenone and 6.00 grams of *m*CPBA. Remember that, as we saw in

Problem 10, the commercially available *m*CPBA is 77% the desired reagent. What is the percent yield if 4.50 grams of *tert*-butyl benzoate are obtained?

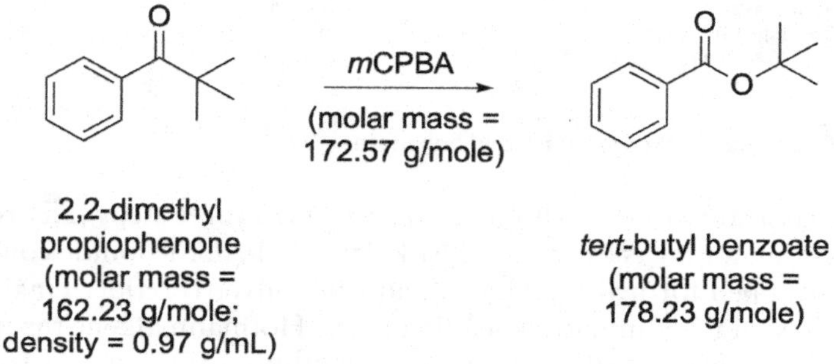

2,2-dimethyl propiophenone
(molar mass = 162.23 g/mole; density = 0.97 g/mL)

*m*CPBA
(molar mass = 172.57 g/mole)

tert-butyl benzoate
(molar mass = 178.23 g/mole)

Figure 9.17 Baeyer–Villiger oxidation of 2,2-dimethylpropiophenone.

18. The reaction of benzoyl chloride with methylamine yields N-methylbenzamide (Figure 9.18). Plan a synthesis of 20.00 grams of N-methylbenzamide. When you conducted this reaction previously, you obtained close to a quantitative (100%) yield, so no yield-based adjustments will be required.

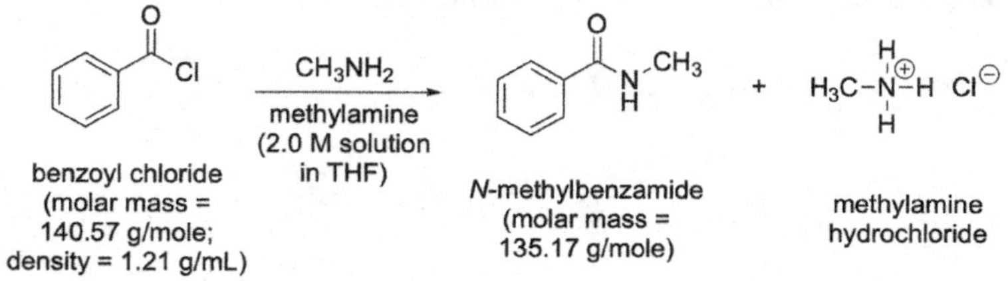

benzoyl chloride
(molar mass = 140.57 g/mole; density = 1.21 g/mL)

CH₃NH₂
methylamine
(2.0 M solution in THF)

N-methylbenzamide
(molar mass = 135.17 g/mole)

methylamine hydrochloride

Figure 9.18 Amide formation from benzoyl chloride.

19. The reaction of methyl benzoate with a Grignard reagent (methylmagnesium bromide, CH_3MgBr) produces the tertiary alcohol 2-phenyl-2-propanol (Figure 9.19). Plan a synthesis of 5.50 grams of 2-phenyl-2-propanol, assuming that the reaction will give approximately a 65% yield. Note that you need not plan for specific amounts of acid or water in step 2 of the sequence; a large excess will be used. Also, note that inorganic by-products have been omitted for simplicity.

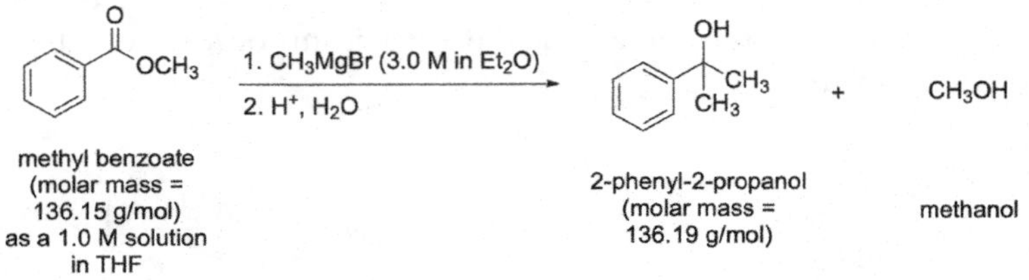

Figure 9.19 Grignard reaction of methyl benzoate.

20. A classical Hofmann elimination consists of three steps (Figure 9.20). First, an amine is exhaustively alkylated. Then, the iodide counterion is swapped for hydroxide using aqueous silver(I) oxide. Finally, heat causes an E2 elimination, leading to the Hofmann alkene through the lower-energy transition state. Plan a synthesis of 7.50 mL of 1-pentene via this sequence, assuming an overall yield of about 90%. Note that undesired by-products have been omitted throughout for simplicity.

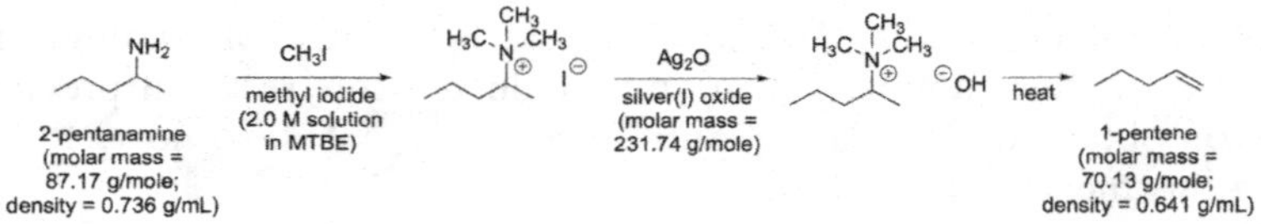

Figure 9.20 Hofmann elimination of 2-pentanamine.

Chapter 10

Solutions to practice problems

1. The chemical equation was not balanced as given. However, when we balance the equation by adding the by-product sodium iodide, we see that this will have no impact on our calculations (Figure 10.1).

   Whenever we are asked to calculate the percent yield, calculation of the theoretical yield is implied because it is the denominator in the percent yield calculation. So, we begin with the determination of theoretical yield.

   The first calculation converts the given volume of cyclohexyl iodide to the mass of product that could be made. Density is used to convert volume cyclohexyl iodide (abbreviated as "CHxI") to mass. Then, the molar mass is used to convert to moles cyclohexyl iodide. The mole ratio is 1:1. Finally, the molar mass of product (abbreviated as "CHxSMe") is used to convert moles of product to mass of product. Since the original given volume contained three significant figures, so does the answer.

$$8.25 \; mL \; CHxI \times \frac{1.624 \; gCHxI}{1 \; mL \; CHxI} \times \frac{1 \; moleCHxI}{210.06 \; gCHxI} \times \frac{1 \; mole \; CHxSMe}{1 \; mole \; CHxI}$$

$$\times \frac{130.25 \; gCHxSMe}{1 \; mole \; CHxSMe} \approx 8.31 gCHxSMe$$

   The second calculation is shorter because we are given a mass of sodium methanethiolate (abbreviated as "MeSNa"). Molar mass is used to convert to moles. Once again, the mole ratio is 1:1. Finally, moles of product are converted to mass using the appropriate molar mass. Both the given mass and the answer contain three significant digits.

$$4.50 \; g \; MeSNa \times \frac{1 \; mole \; MeSNa}{70.08 \; g \; MeSNa} \times \frac{1 \; mole \; CHxSMe}{1 \; mole \; MeSNa} \times \frac{130.25 \; g \; CHxSMe}{1 \; mole \; CHxSMe}$$

$$\approx 8.36 \; g \; CHxSMe$$

DOI: 10.1201/9781003587408-10

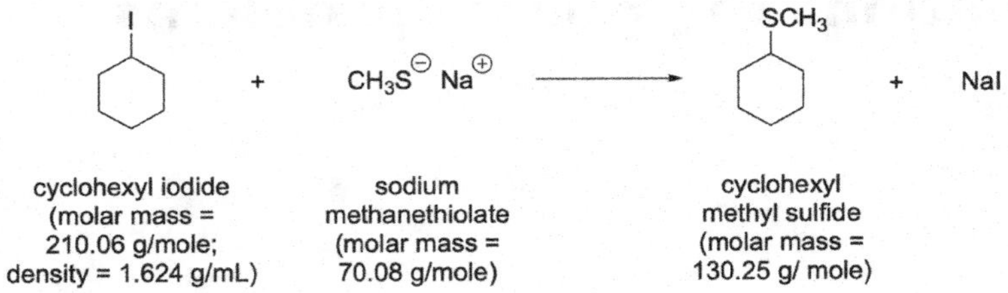

cyclohexyl iodide
(molar mass =
210.06 g/mole;
density = 1.624 g/mL)

sodium
methanethiolate
(molar mass =
70.08 g/mole)

cyclohexyl
methyl sulfide
(molar mass =
130.25 g/ mole)

Figure 10.1 Balanced S$_N$2 reaction between cyclohexyl iodide and sodium methanethiolate.

The smaller value, 8.31 grams, is the theoretical yield of cyclohexyl methyl sulfide, making cyclohexyl iodide the limiting reactant. Knowing that we obtained 6.99 grams of product, we can now calculate the percent yield. Since both the amount of product obtained and the theoretical yield contain three significant figures, so does the answer.

$$Percent\ Yield = \frac{amount\ of\ product\ obtained}{theoretical\ yield} \times 100$$

$$Percent\ Yield = \frac{6.99\ \cancel{g\ CHxSMe}}{8.31\ \cancel{g\ CHxSMe}} \times 100 \approx 84.1\%$$

2. Although this chemical equation was not balanced as provided, the balanced form simply includes two overlooked by-products (potassium iodide and *tert*-butyl alcohol) and does not, therefore, impact the calculations (Figure 10.2).

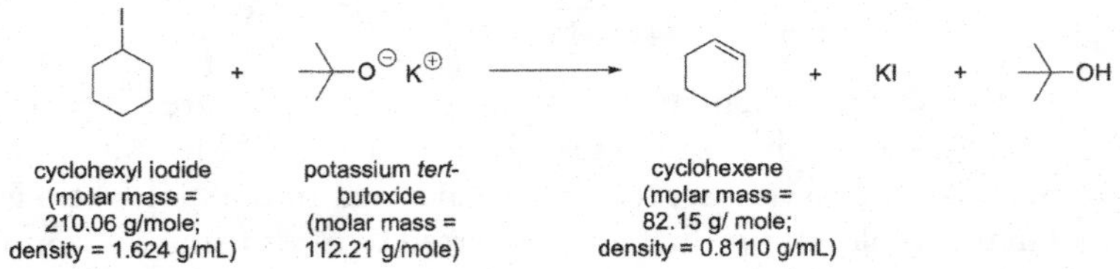

cyclohexyl iodide
(molar mass =
210.06 g/mole;
density = 1.624 g/mL)

potassium *tert*-
butoxide
(molar mass =
112.21 g/mole)

cyclohexene
(molar mass =
82.15 g/ mole;
density = 0.8110 g/mL)

Figure 10.2 Balanced E2 reaction between cyclohexyl iodide and potassium *tert*-butoxide.

The calculation of percent yield must be preceded by determination of the theoretical yield.

The first calculation begins with the given volume of cyclohexyl iodide (abbreviated as "CHxI"). Density is used to convert this to a mass, and molar mass is used to convert to moles cyclohexyl iodide. The mole ratio is 1:1. Then, moles of cyclohexene (abbreviated as "ch") are converted to mass using molar mass and finally to a volume using density. Remember to retain the same number of significant digits (three) as in the original given quantity.

$$8.25 \; mLCHxI \times \frac{1.624 \; g\,CHxI}{1 \; mL\,CHxI} \times \frac{1 \; mole\,CHxI}{210.06 \; g\,CHxI} \times \frac{1 \; mole\,ch}{1 \; mole\,CHxI}$$

$$\times \frac{82.15 \; g\,ch}{1 \; mole\,ch} \times \frac{1 \; mL \; ch}{0.8110 \; g\,ch} \approx 6.46 \; mL \; cyclohexene$$

The second calculation begins with the mass of potassium *tert*-butoxide (abbreviated as "KOtBu"). Molar mass enables conversion to moles. Once again, the mole ratio is merely 1:1. Then, as above, we'll convert moles product to mass and finally volume, using molar mass and density, respectively. As above, remember to retain the same number of significant digits (three) as in the original given quantity.

$$7.25 \; g\,KOtBu \times \frac{1 \; mole\,KOtBu}{112.21 \; g\,KOtBu} \times \frac{1 \; mole\,ch}{1 \; mole\,KOtBu} \times \frac{82.15 \; g\,ch}{1 \; mole\,ch}$$

$$\times \frac{1 \; mL \; ch}{0.8110 \; g\,ch} \approx 6.54 \; mL \; cyclohexene$$

The smaller quantity of product, 6.46 mL, is the theoretical yield. Though we were not asked to state the limiting reactant, it is cyclohexyl iodide. With the theoretical yield in hand, we can now determine the percent yield when 6.12 mL of cyclohexene are obtained. Since each quantity contains three significant digits, so should our answer.

$$Percent \; Yield = \frac{amount \; of \; product \; obtained}{theoretical \; yield} \times 100$$

$$Percent \; Yield = \frac{6.12 \; mLcyclohexene}{6.46 \; mLcyclohexene} \times 100 \approx 94.7\%$$

3. Checking to make sure that the reaction equation is balanced is the first order of business. A reaction like this, with major and minor products, can be confusing. In particular, you must think very clearly about the mole ratio. Our hint was that one molecule of 2-pentanol can yield only one molecule of alkene, regardless of which one it is. Whatever we can say with regard to molecules can also be said with regard to moles. So, one mole of 2-pentanol can yield only 1 mole of alkene. The fact that there are options for the structure of alkene

produced is actually irrelevant as far as the calculation of theoretical yield goes. This may be easier to see if we remove the minor products from the reaction equation. This highlights the fact that the reaction is balanced once we add in the overlooked by-product, water. Whether the alkene produced is *trans*-2-pentene, *cis*-2-pentene, 1-pentene, or a mixture of all three has no bearing on the balancing of the equation or the subsequent calculations (Figure 10.3).

Now, we are prepared to address the theoretical yield calculation. In this case, our task is made easier by the fact that there is only one reactant, 2-pentanol. Remember that catalysts are neither created nor destroyed in chemical reactions and are, therefore, neither reactants nor products. We begin with the given volume of 2-pentanol (abbreviated as "p") and use its density followed by molar mass to convert to moles. The mole ratio is 1:1. Lastly, the moles of *trans*-2-pentene (abbreviated as "t2p") are converted to grams using its molar mass. Since the original given mass contains four significant figures, the answer has four significant digits too.

$$40.00 \;\cancel{mL\;p} \times \frac{0.812 \;\cancel{g\;p}}{1\;\cancel{mL\;p}} \times \frac{1\;\cancel{mole\;p}}{88.15\;\cancel{g\;p}} \times \frac{1\;\cancel{mole\;t2p}}{1\;\cancel{mole\;p}}$$

$$\times \frac{70.13\;g\;t2p}{1\;\cancel{mole\;t2p}} \approx 25.84 \; g \; trans-2-pentene$$

The theoretical yield is 25.84 grams. How this mass is distributed across the three possible products is inconsequential for this problem. We could make at most 25.84 grams of *trans*-2-pentene *or* 25.84 grams of *cis*-2-pentene *or* 25.84 grams of 1-pentene *or* 25.84 grams of a mixture of all three. The last possibility, making as much as 25.84 grams of a mixture of all three alkenes, is what actually happens in the laboratory, so that's how we'll phrase things for the percent yield calculation.

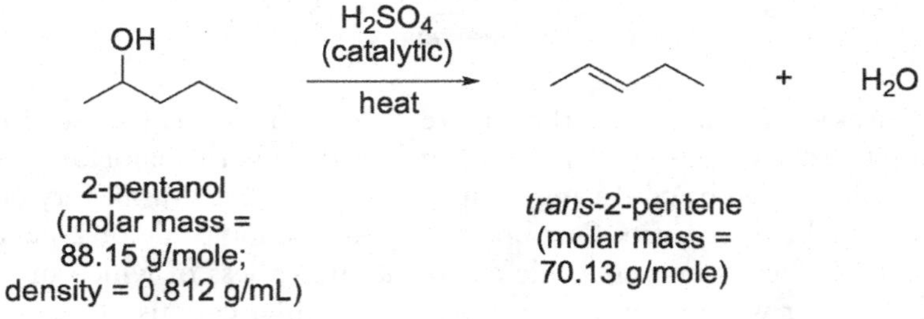

Figure 10.3 Simplified dehydration of 2-pentanol.

$$Percent\,Yield = \frac{amount\,of\,product\,obtained}{theoretical\,yield} \times 100$$

$$Percent\,Yield = \frac{15.2 \; \cancel{g\,alkenes\,mixture}}{25.84 \; \cancel{g\,alkenes\,mixture}} \times 100 \approx 58.8\%$$

The denominator has four significant figures, but the numerator has only three, which limits the answer to three significant digits.

4. This reaction equation is balanced as given, so we can proceed directly to the determination of limiting reactant and calculation of theoretical yield.

The initial calculation begins with the given mass of 2-methyl-1-cyclopentanol (abbreviated as "mcp") and uses its molar mass to convert to moles. The mole ratio is simply 1:1. Finally, the moles of 1-bromo-2-methylcyclopentane product (abbreviated as "bmcp") are converted to grams using its molar mass. Three significant figures appear in both the given quantity and the answer.

$$2.25 \; \cancel{g\,mcp} \times \frac{1 \; \cancel{mole\,mcp}}{100.16 \; \cancel{g\,mcp}} \times \frac{1 \; \cancel{mole\,bmcp}}{1 \; \cancel{mole\,mcp}} \times \frac{163.06\,g\,bmcp}{1 \; \cancel{mole\,bmcp}}$$

$$\approx 3.66\,g\,bromomethylcyclopentane$$

The second calculation begins with a volume of phosphorus tribromide (PBr_3). Its density and molar mass are used to obtain the moles of PBr_3. Again, the mole ratio is merely 1:1, and the moles of product are converted to mass using its molar mass. Three significant digits are in both the given volume and the final answer.

$$2.25 \; \cancel{mL\,PBr_3} \times \frac{2.852 \; \cancel{g\,PBr_3}}{1 \; \cancel{mL\,PBr_3}} \times \frac{1 \; \cancel{mole\,PBr_3}}{270.69 \; \cancel{g\,PBr_3}} \times \frac{1 \; \cancel{mole\,bmcp}}{1 \; \cancel{mole\,PBr_3}}$$

$$\times \frac{163.06\,g\,bmcp}{1 \; \cancel{mole\,bmcp}} \approx 3.87\,g\,bromomethylcyclopentane$$

The smaller of the two values, 3.66 grams, is the theoretical yield of bromomethylcyclopentane product, showing that methylcyclopentanol is the limiting reactant. The percent yield, when 3.11 grams of product are obtained, can then be calculated.

$$Percent\,Yield = \frac{amount\,of\,product\,obtained}{theoretical\,yield} \times 100$$

$$= \frac{3.11 \; \cancel{g\,bmcp}}{3.66 \; \cancel{g\,bmcp}} \times 100 \approx 85.0\%$$

5. We've been told that the equation is balanced, so we can leap directly into the determination of limiting reactant and theoretical yield. This will require three calculations: one for (S)-1-phenylethanol, sodium hydride, and propyl chloride.

The first calculation begins with the given volume of (S)-1-phenylethanol (abbreviated as "PhEtOH"). Its density is used to convert volume to mass, and its molar mass is used to convert mass to moles. The mole ratio is a straightforward 1:1. Lastly, moles of (S)-(1-propoxyethyl)benzene product (abbreviated as "peb") are converted to a mass using its molar mass. There are three significant figures in the given volume of reactant and the mass of product.

$$5.00 \; \cancel{mL \; PhEtOH} \times \frac{1.012 \; \cancel{g \; PhEtOH}}{1 \; \cancel{mL \; PhEtOH}} \times \frac{1 \; \cancel{mole \; PhEtOH}}{122.17 \; \cancel{g \; PhEtOH}}$$

$$\times \frac{1 \; \cancel{mole \; peb}}{1 \; \cancel{mole \; PhEtOH}} \times \frac{164.25 \; g \; peb}{1 \; \cancel{mole \; peb}}$$

$$\approx 6.80 \; g \, (\text{propoxyethyl}) \text{benzene}$$

The second calculation begins with the given mass of sodium hydride (NaH). Its molar mass is used to ascertain moles. Again, the mole ratio is 1:1. And, the moles of product are converted to grams using molar mass. This time there are four significant digits in the given mass, so the product mass has four as well.

$$1.000 \; \cancel{g \; NaH} \times \frac{1 \; \cancel{mole \; NaH}}{24.00 \; \cancel{g \; NaH}} \times \frac{1 \; \cancel{mole \; peb}}{1 \; \cancel{mole \; NaH}}$$

$$\times \frac{164.25 \; g \; peb}{1 \; \cancel{mole \; peb}} \approx 6.844 \; g \, (\text{propoxyethyl}) \text{benzene}$$

The third calculation starts with the given volume of propyl chloride (PrCl). Its density and molar mass are used to effect conversion to moles. The 1:1 mole ratio is employed again, and the moles of product are convert to grams using its molar mass. There are three significant figures in the given volume and calculated mass.

$$3.70 \; \cancel{mL \; PrCl} \times \frac{0.890 \; \cancel{g \; PrCl}}{1 \; \cancel{mL \; PrCl}} \times \frac{1 \; \cancel{mole \; PrCl}}{78.54 \; \cancel{g \; PrCl}} \times \frac{1 \; \cancel{mole \; peb}}{1 \; \cancel{mole \; PrCl}}$$

$$\times \frac{164.25 \; g \; peb}{1 \; \cancel{mole \; peb}} \approx 6.89 \; g \, (\text{propoxyethyl}) \text{benzene}$$

Comparing all three outcomes shows the smallest value to be 6.80 grams of (S)-(1-propoxyethyl)benzene. This is the theoretical yield, and it was obtained from (S)-1-phenylethanol, which is therefore the limiting reactant.

Now, we may calculate the percent yield when 6.251 grams of product are obtained. Though the numerator contains four significant figures, the denominator has only three, so the answer contains three.

$$Percent\ Yield = \frac{amount\ of\ product\ obtained}{theoretical\ yield} \times 100$$

$$= \frac{6.251\ \cancel{g\ peb}}{6.80\ \cancel{g\ peb}} \times 100 \approx 91.9\%$$

6. This reaction equation is balanced as written.

The calculation of percent yield first necessitates the determination of theoretical yield, which requires two calculations. The first calculation employs the given volume of (S)-styrene oxide (abbreviated as "so"). Its density can be used to convert volume to mass. Then, its molar mass can be used to convert mass to moles. The mole ratio is 1:1. The final manipulation is the conversion of moles of product (abbreviated as "dmape") to mass of product using its molar mass. There are three significant figures in both the starting and ending values.

$$6.75\ \cancel{mL\ so} \times \frac{1.051\ \cancel{g\ so}}{1\ \cancel{mL\ so}} \times \frac{1\ \cancel{mole\ so}}{120.15\ \cancel{g\ so}} \times \frac{1\ \cancel{mole\ dmape}}{1\ \cancel{mole\ so}}$$

$$\times \frac{165.24\ g\ dmape}{1\ \cancel{mole\ dmape}} \approx 9.76\ g\,(\text{dimethylamino})\text{phenylethanol}$$

The second calculation is directly analogous. The steps are identical, but of course the volume, density, and molar mass of dimethylamine ("dma") are used instead. Again, we find three significant digits at the beginning and end of the calculation.

$$4.00\ \cancel{mL\ dma} \times \frac{0.680\ \cancel{g\ dma}}{1\ \cancel{mL\ dma}} \times \frac{1\ \cancel{mole\ dma}}{45.09\ \cancel{g\ dma}} \times \frac{1\ \cancel{mole\ dmape}}{1\ \cancel{mole\ dma}}$$

$$\times \frac{165.24\ g\ dmape}{1\ \cancel{mole\ dmape}} \approx 9.97\ g\,(\text{dimethylamino})\text{phenylethanol}$$

The smaller of the two values is the theoretical yield: 9.76 grams of product. This shows that (S)-styrene oxide is the limiting reactant.

The percent yield, when 8.76 grams of (S)-2-(dimethylamino)-1-phenylethanol are obtained, can now be determined.

$$Percent\ Yield = \frac{amount\ of\ product\ obtained}{theoretical\ yield} \times 100$$

$$= \frac{8.76\ \text{g dmape}}{9.76\ \text{g dmape}} \times 100 \approx 89.8\%$$

7. Did you try to solve the problem without first balancing the equation? If so, go back, balance the equation, and adjust your calculations accordingly before proceeding.

Balancing the equation is achieved by placing a stoichiometric coefficient of 2 in front of HCl (Figure 10.4).

Now, we can begin our theoretical yield calculations, which are necessary for the eventual determination of percent yield. The first calculation starts with the given mass of 1,3-dimethylenecyclohexane (abbreviated as "dmch"). Its molar mass is used to convert mass to moles. The mole ratio in this case is 1:1. Then, the moles of 1,3-dichloro-1,3-dimethylcyclohexane (abbreviated as "dcdmch") are converted to grams using its molar mass. There are four significant figures in the given mass of reactant and the calculated mass of product.

$$11.50\ \text{g dmch} \times \frac{1\ \text{mole dmch}}{108.18\ \text{g dmch}} \times \frac{1\ \text{mole dcdmch}}{1\ \text{mole dmch}}$$

$$\times \frac{181.10\ \text{g dcdmch}}{1\ \text{mole dcdmch}} \approx 19.25\ g\ dcdmch$$

The second calculation begins with the given volume of HCl solution. Since this volume is measured in milliliters, we must first convert to liters. Then, we can use the molarity of the solution (4.0 M = 4.0 moles solute per liter of solution) to obtain the moles of HCl. This time, the stoichiometric ratio is 2 moles HCl:1 mole product. Finally, the molar mass of product enables conversion of moles of product to

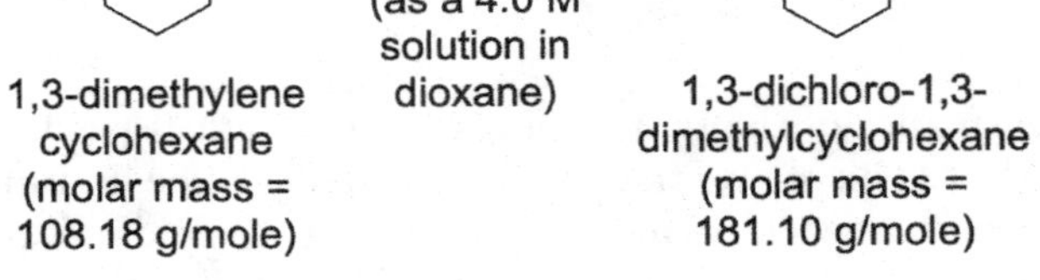

1,3-dimethylene cyclohexane (molar mass = 108.18 g/mole)

2 HCl
(as a 4.0 M solution in dioxane)

1,3-dichloro-1,3-dimethylcyclohexane (molar mass = 181.10 g/mole)

Figure 10.4 A balanced ionic hydrohalogenation.

mass of product. In this instance, there are three significant digits in both the given volume of reactant and the calculated mass of product.

$$54.0 \; \cancel{mL \; HCl \; solution} \times \frac{1 \; \cancel{L}}{1000 \; \cancel{mL}} \times \frac{4.0 \; \cancel{moles \; HCl}}{1 \; \cancel{L \; HCl \; solution}} \times \frac{1 \; \cancel{mole \; dcdmch}}{2 \; \cancel{mole \; HCl}}$$

$$\times \frac{181.10 \; g \; dcdmch}{1 \; \cancel{mole \; dcdmch}} \approx 19.6 \; g \; dcdmch$$

The smaller of the two calculated values, 19.25 grams, is the theoretical yield, and 1,3-dimethylenecyclohexane is the limiting reactant. We can now determine the percent yield for the reaction when 15.3 grams of 1,3-dichloro-1,3-dimethylcyclohexane are generated. Though the denominator contains four significant figures, the numerator has only three, so the answer must have three as well.

$$Percent \; Yield = \frac{amount \; of \; product \; obtained}{theoretical \; yield} \times 100$$

$$= \frac{15.3 \; \cancel{g \; dcdmch}}{19.25 \; \cancel{g \; dcdmch}} \times 100 \approx 79.5\%$$

8. (a) We've been told that the 33% solution of HBr in acetic acid (HOAc) is a weight percentage, meaning it is the ratio of solute weight to total solution weight multiplied by 100.

$$33\% = \frac{weight \; HBr}{weight \; HBr \; solution} \times 100$$

Therefore, in 1 kg of HBr solution, there would be 0.330 kg (or 330 grams) of HBr.

$$33\% = \frac{weight \; HBr}{1 \; kg} \times 100$$

$$\frac{33\% \times (1 \; kg)}{100} = weight \; HBr = 0.33 \; kg$$

$$0.33 \; \cancel{kg} \times \frac{1000 \; g}{1 \; \cancel{kg}} = 330g$$

To summarize, the 33% w/w solution has 330 grams of HBr in every 1,000 grams of HBr solution.

$$\frac{330 \ g \ HBr}{1000 \ g \ HBr \ solution}$$

(b) The equation is balanced as written, so we can dive right into the theoretical yield calculations. The first one begins with the given volume of 2-bromostyrene (abbreviated as "bs"). Its density and molar mass can be used to convert the volume to moles. The mole ratio is 1:1. Then, the molar mass of 1-bromo-2-(2-bromoethyl) benzene (abbreviated as "bbeb") can be used to obtain the mass of product. There are three significant figures in the given volume of reactant and the calculated mass of product.

$$5.00 \ \cancel{mL \ bs} \times \frac{1.460 \ \cancel{g \ bs}}{1 \ \cancel{mL \ bs}} \times \frac{1 \ \cancel{mole \ bs}}{183.05 \ \cancel{g \ bs}} \times \frac{1 \ \cancel{mole \ bbeb}}{1 \ \cancel{mole \ bs}} \times \frac{263.96 \ g \ bbeb}{1 \ \cancel{mole \ bbeb}}$$
$$\approx 10.5 \ g \ bbeb$$

The second calculation is slightly more involved, and we'll need to use the conversion factor derived in part (a). We begin with the given volume of HBr solution (abbreviated as "HBr soln"). The density of the solution can be used to convert this volume to a mass of the solution. Then, we need to use our new conversion factor (330 grams of HBr/1,000 grams of HBr solution) to convert mass of the solution to mass of HBr. Once we have the mass of HBr, its molar mass enables conversion to moles. The mole ratio is 1:1, and the molar mass of product allows us to convert moles of product to grams of product. There are four significant digits in the given volume of HBr solution and in the calculated mass of product.

$$10.00 \ \cancel{mL \ HBr \ soln} \times \frac{1.40 \ \cancel{g \ HBr \ soln}}{1 \ \cancel{mL \ HBr \ soln}} \times \frac{330 \ \cancel{g \ HBr}}{1000 \ \cancel{g \ HBr \ soln}} \times \frac{1 \ \cancel{mole \ HBr}}{80.91 \ \cancel{g \ HBr}}$$
$$\times \frac{1 \ \cancel{mole \ bbeb}}{1 \ \cancel{mole \ HBr}} \times \frac{263.96 \ g \ bbeb}{1 \ \cancel{mole \ bbeb}} \approx 15.07 \ g \ bbeb$$

The smaller of the two product masses, 10.5 grams, is the theoretical yield, and it is derived from 2-bromostyrene, making that the limiting reactant.

(c) Now that we know the theoretical yield, we can calculate the percent yield for the reaction if 8.67 grams of product are obtained.

$$Percent \ Yield = \frac{amount \ of \ product \ obtained}{theoretical \ yield} \times 100$$
$$= \frac{8.67 \ \cancel{g \ bbeb}}{10.5 \ \cancel{g \ bbeb}} \times 100 \approx 82.6\%$$

9. The reaction equation is balanced as written. An important note is that dichloromethane (CH_2Cl_2) is an inert solvent for the reaction. Inert solvents do not play a role in the chemical transformation. They are merely present to dissolve the reactants and enable them to react with one another. Therefore, we can ignore the solvent when balancing the equation and when determining theoretical yield.

Both of the calculations for determination of theoretical yield are very similar. In each case, we are beginning with a neat liquid reactant (2-methyl-2-butene is abbreviated as "mb"), so its density is used to convert volume to mass. Then, its molar mass is used to convert mass to moles. In each case, the mole ratio of reactant to product is 1:1. Finally, moles of product (abbreviated as "dbmb") are converted to mass using its molar mass. Both given volumes have four significant digits, so the calculated product masses will as well.

$$20.00 \; mL \; mb \times \frac{0.662 \; g \; mb}{1 \; mL \; mb} \times \frac{1 \; mole \; mb}{70.13 \; g \; mb} \times \frac{1 \; mole \; dbmb}{1 \; mole \; mb}$$

$$\times \frac{229.94 \; g \; dbmb}{1 \; mole \; dbmb} \approx 43.41 \; g \; dbmb$$

$$10.00 \; mL \; Br_2 \times \frac{3.12 \; g \; Br_2}{1 \; mL \; Br_2} \times \frac{1 \; mole \; Br_2}{159.81 \; g \; Br_2} \times \frac{1 \; mole \; dbmb}{1 \; mole \; Br_2}$$

$$\times \frac{229.94 \; g \; dbmb}{1 \; mole \; dbmb} \approx 44.89 \; g \; dbmb$$

The smaller of the two values is 43.41 grams of 2,3-dibromo-2-methylbutane. This is the theoretical yield, and it was the 2-methyl-2-butene that limited the amount of product that could be formed. The percent yield, when 35.31 grams of product are isolated, can now be determined. Since both the numerator and denominator have four significant figures, the answer will too.

$$Percent \; Yield = \frac{amount \; of \; product \; obtained}{theoretical \; yield} \times 100$$

$$= \frac{35.31 \; g \; dbmb}{43.41 \; g \; dbmb} \times 100 \approx 81.34\%$$

10. (a) Recall that, when planning syntheses, you are essentially conducting theoretical yield calculations in reverse. Our two calculations will begin with the desired volume of styrene oxide (abbreviated as "so"). Its density is used to convert volume to mass. Then, its molar mass is used to obtain moles. In each case, the stoichiometric ratio

is 1:1. In the first calculation, the moles of styrene (abbreviated as "s") are converted to mass using its molar mass and then to volume using its density.

$$10.00 \; \cancel{mL \; so} \times \frac{1.054 \; \cancel{g \; so}}{1 \; \cancel{mL \; so}} \times \frac{1 \; \cancel{mole \; so}}{120.15 \; \cancel{g \; so}} \times \frac{1 \; \cancel{mole \; s}}{1 \; \cancel{mole \; so}}$$

$$\times \frac{104.15 \; \cancel{g \; s}}{1 \; \cancel{mole \; s}} \times \frac{1 \; mL \; s}{0.906 \; \cancel{g \; s}} \approx 10.08 \; mL \; styrene$$

In the second calculation, the moles of *m*CPBA are converted to a mass using its molar mass.

$$10.00 \; \cancel{mL \; so} \times \frac{1.054 \; \cancel{g \; so}}{1 \; \cancel{mL \; so}} \times \frac{1 \; \cancel{mole \; so}}{120.15 \; \cancel{g \; so}} \times \frac{1 \; \cancel{mole \; mCPBA}}{1 \; \cancel{mole \; so}}$$

$$\times \frac{172.57 \; g \; mCPBA}{1 \; \cancel{mole \; mCPBA}} \approx 15.14 \; g \; mCPBA$$

So, we'd need 10.08 mL of styrene and 15.14 grams of *m*CPBA in order to prepare about 10.00 mL of styrene oxide.

(b) Sometimes reagent samples do not actually contain 100% of the desired substance. This is the case with "technical grade" reagents. If the commercially acquired *m*CPBA is only 77% *m*CPBA, we must simply divide the needed amount by 0.77 to adjust the quantity upward and account for this.

$$\frac{15.14 \; g \; mCPBA}{0.77} = 19.66 \; g \; mCPBA$$

We'll need to use 19.66 grams of this sample, which is only 77% *m*CPBA, in order to actually deliver the needed 15.14 grams of *m*CPBA to the reaction.

(c) If in our hands the reaction tends to yield only about 85% yield, we should divide the quantity of each reactant by 0.85 to ensure that we make our target: roughly 10.00 mL of styrene oxide.

$$\frac{10.08 \; mL \; styrene}{0.85} = 11.86 \; mL \; styrene$$

$$\frac{19.66 \; g \; mCPBA}{0.85} = 23.13 \; g \; mCPBA$$

At this point, we've done everything necessary to answer the question, but just in case you aren't totally convinced that this is all correct, we can prove it to ourselves by conducting the theoretical yield calculations. The first begins with the 11.86 mL of styrene that we determined we'd need. We use its density and then its molar mass to convert to moles. The mole ratio is 1:1. Next, moles of styrene oxide are converted to a mass and then a volume, using its molar mass and density, respectively.

$$11.86 \; \cancel{mL\,s} \times \frac{0.906 \; \cancel{g\,s}}{1 \; \cancel{mL\,s}} \times \frac{1 \; \cancel{mole\,s}}{104.15 \; \cancel{g\,s}} \times \frac{1 \; \cancel{mole\,so}}{1 \; \cancel{mole\,s}} \times \frac{120.15 \; \cancel{g\,so}}{1 \; \cancel{mole\,so}}$$

$$\times \frac{1 \; mL \; so}{1.054 \; \cancel{g\,so}} \approx 11.76 \; mL \; styrene \; oxide$$

In the other calculation, the mass of *m*CPBA we decided to use is first multiplied by 0.77 to account for the fact that only 77% of the sample is actually *m*CPBA. Then, molar mass is used to convert this mass to moles. The stoichiometric ratio is again 1:1. We finish this calculation as we did the former one, using styrene oxide's molar mass and density to convert moles to mass and then to volume.

$$23.13 \; \cancel{g\,mCPBA} \times 0.77 \times \frac{1 \; \cancel{mole\,mCPBA}}{172.57 \; \cancel{g\,mCPBA}} \times \frac{1 \; \cancel{mole\,so}}{1 \; \cancel{mole\,mCPBA}}$$

$$\times \frac{120.15 \; \cancel{g\,so}}{1 \; \cancel{mole\,so}} \times \frac{1 \; mL \; so}{1.054 \; \cancel{g\,so}} \approx 11.76 \; mL \; styrene \; oxide$$

The two amounts match perfectly, and our theoretical yield is 11.76 mL. If it seems like we've prepared too much styrene oxide, remember that we still need to account for the fact that the reaction tends to give us just 85% yield.

$$Percent \; Yield = \frac{amount \; of \; product \; obtained}{theoretical \; yield} \times 100$$

$$85\% = \frac{amount \; of \; product \; obtained}{11.76 \; mL \; styrene \; oxide} \times 100$$

$$\frac{85\% \times \left(11.76 \; mL \; styrene \; oxide\right)}{100} = amount \; of \; product \; obtained$$

$$\approx 10.00 \; mL \; styrene \; oxide$$

After doing so, it is clear that we can anticipate making exactly the desired amount of styrene oxide, about 10.00 mL.

11. (a) Planning our synthesis will require three calculations to determine the amounts of styrene, molecular bromine, and sodamide needed.

All three calculations will begin the same way: The desired volume of phenylacetylene (abbreviated as "pa") is converted to a mass and then to moles, using its density and molar mass, respectively. For the first two calculations, the mole ratio is 1:1. Moles of the desired reactant are then converted to grams followed by milliliters, using the appropriate molar mass and density in sequence.

$$15.00 \; \cancel{mL \; pa} \times \frac{0.930 \; \cancel{g \; pa}}{1 \; \cancel{mL \; pa}} \times \frac{1 \; \cancel{mole \; pa}}{102.14 \; \cancel{g \; pa}} \times \frac{1 \; \cancel{mole \; s}}{1 \; \cancel{mole \; pa}} \times \frac{104.15 \; \cancel{g \; s}}{1 \; \cancel{mole \; s}}$$

$$\times \frac{1 \; mL \; s}{0.906 \; \cancel{g \; s}} \approx 15.70 \; mL \; styrene$$

$$15.00 \; \cancel{mL \; pa} \times \frac{0.930 \; \cancel{g \; pa}}{1 \; \cancel{mL \; pa}} \times \frac{1 \; \cancel{mole \; pa}}{102.14 \; \cancel{g \; pa}} \times \frac{1 \; \cancel{mole \; Br_2}}{1 \; \cancel{mole \; pa}} \times \frac{159.81 \; \cancel{g \; Br_2}}{1 \; \cancel{mole \; Br_2}}$$

$$\times \frac{1 \; mL \; Br_2}{3.12 \; \cancel{g \; Br_2}} \approx 6.996 \; mL \; Br_2$$

For the third calculation, we are told to use 3.1 molar equivalents of sodamide. Finally, the moles of sodamide are converted to a mass using its molar mass. In all three calculations, the four significant figures in the desired volume of phenylacetylene are maintained in the calculated quantity of reactant.

$$15.00 \; \cancel{mL \; pa} \times \frac{0.930 \; \cancel{g \; pa}}{1 \; \cancel{mL \; pa}} \times \frac{1 \; \cancel{mole \; pa}}{102.14 \; \cancel{g \; pa}} \times \frac{3.1 \; \cancel{mole \; NaNH_2}}{1 \; \cancel{mole \; pa}}$$

$$\times \frac{39.01 \; g \; NaNH_2}{1 \; \cancel{mole \; NaNH_2}} \approx 16.52 \; g \; NaNH_2$$

(b) If we are aware that the reaction will give us about 90% yield, then we can scale up the amount of each reactant to be used by dividing by 0.90.

$$\frac{15.70 \; mL \; styrene}{0.90} = 17.44 \; mL \; styrene$$

$$\frac{6.996 \; mL \; Br_2}{0.90} = 7.773 \; mL \; Br_2$$

$$\frac{16.52 \; g \; sodamide}{0.90} = 18.36 \; g \; sodamide$$

12. (a) We'll need three calculations to plan this synthesis: for the amounts of phenylacetylene (abbreviated as "pa"), sodamide ($NaNH_2$),

and butyl bromide (abbreviated as "bb") to use. In each case, we begin with the desired mass of butylphenylacetylene (abbreviated as "bpa") and use its molar mass to convert to moles. The mole ratios are all 1:1. Then, the calculations are completed by using the reactant's molar mass to convert moles to grams (in the case of sodamide) or by doing this and then using density to convert mass to volume (in the cases of phenylacetylene and butyl bromide). In each instance, four significant figures are maintained in the calculated values.

$$25.00 \; \cancel{g \, bpa} \times \frac{1 \; \cancel{mole \, bpa}}{158.24 \; \cancel{g \, bpa}} \times \frac{1 \; \cancel{mole \, pa}}{1 \; \cancel{mole \, bpa}} \times \frac{102.14 \; \cancel{g \, pa}}{1 \; \cancel{mole \, pa}}$$

$$\times \frac{1 \; mL \, pa}{0.930 \; \cancel{g \, pa}} \approx 17.35 \; mL \; phenylacetylene$$

$$25.00 \; \cancel{g \, bpa} \times \frac{1 \; \cancel{mole \, bpa}}{158.24 \; \cancel{g \, bpa}} \times \frac{1 \; \cancel{mole \, NaNH_2}}{1 \; \cancel{mole \, bpa}}$$

$$\times \frac{39.01 \; g \, NaNH_2}{1 \; \cancel{mole \, NaNH_2}} \approx 6.163 \; g \; NaNH_2$$

$$25.00 \; \cancel{g \, bpa} \times \frac{1 \; \cancel{mole \, bpa}}{158.24 \; \cancel{g \, bpa}} \times \frac{1 \; \cancel{mole \, bb}}{1 \; \cancel{mole \, bpa}} \times \frac{137.02 \; \cancel{g \, bb}}{1 \; \cancel{mole \, bb}}$$

$$\times \frac{1 \; mL \, bb}{1.27 \; \cancel{g \, bb}} \approx 17.05 \; mL \; butyl \; bromide$$

(b) If we can anticipate a 75% yield, then each quantity of reactant should be scaled up through division by 0.75.

$$\frac{17.35 \; mL \; phenylacetylene}{0.75} = 23.13 \; mL \; phenylacetylene$$

$$\frac{6.163 \; g \; sodamide}{0.75} = 8.217 \; g \; sodamide$$

$$\frac{17.05 \; mL \; butyl \; bromide}{0.75} = 22.73 \; mL \; butyl \; bromide$$

13. (a) If we label the carbon atoms in 1,3-butadiene as a–d and the carbons in dimethylacetylene as e–h, it helps us to see that the Diels–Alder adduct contains two units derived from 1,3-butadiene and one unit derived from dimethylacetylene (Figure 10.5). Therefore,

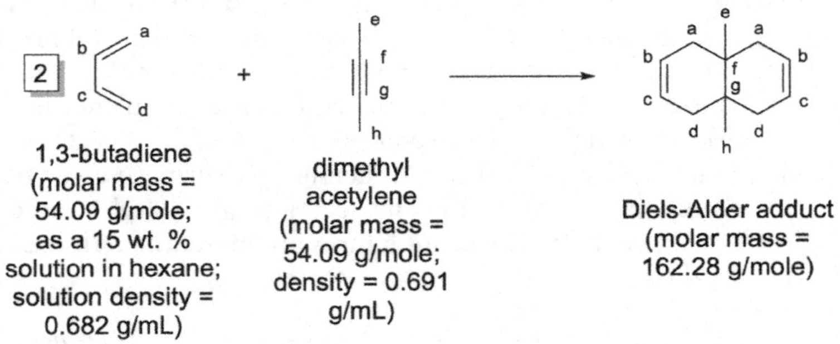

1,3-butadiene (molar mass = 54.09 g/mole; as a 15 wt. % solution in hexane; solution density = 0.682 g/mL)	dimethyl acetylene (molar mass = 54.09 g/mole; density = 0.691 g/mL)	Diels-Alder adduct (molar mass = 162.28 g/mole)

Figure 10.5 Balanced Diels–Alder reaction between 1,3-butadiene and dimethylacetylene.

we must place a stoichiometric coefficient of 2 in front of 1,3-buta-diene to balance the equation.

(b) We've been told that the 15% solution of 1,3-butadiene in hexane is a weight percentage, meaning it is the ratio of solute weight to total solution weight multiplied by 100.

$$15\% = \frac{weight\ 1,3-butadiene}{weight\ 1,3-butadiene\ solution} \times 100$$

Therefore, in 1 kg of 1,3-butadiene solution, there would be 0.150 kg (or 150 g) 1,3-butadiene.

$$15\% = \frac{weight\ 1,3-butadiene}{1\ kg} \times 100$$

$$\frac{15\% \times (1\ kg)}{100} = weight\ 1,3-butadiene = 0.15\ kg$$

$$0.15\ \cancel{kg} \times \frac{1000\ g}{1\ \cancel{kg}} = 150\ g$$

To summarize, the 15% w/w solution has 150 grams of 1,3-butadiene in every 1,000 grams of 1,3-butadiene solution.

$$\frac{150\ g\ 1,3-butadiene}{1000\ g\ 1,3-butadiene\ solution}$$

(c) Planning the synthesis of 5.00 grams of Diels–Alder adduct requires two calculations. In both, we begin by dividing the desired mass of Diels–Alder adduct (abbreviated as "adduct") by its molar mass to obtain moles. In the calculation of the required amount of 1,3-butadiene solution (abbreviated as "bd soln"), the mole ratio is 1 mole adduct:2 moles 1,3-butadiene (abbreviated as "bd"). Then, its molar mass is used to convert moles of 1,3-butadiene to grams of 1,3-butadiene. Now, we can employ the new conversion factor derived in part (b) above (150 grams of 1,3-butadiene/1,000 grams of 1,3-butadiene solution) to convert mass of 1,3-butadiene to mass of the solution. Finally, the solution's density is used to convert mass to volume.

$$5.00 \;\cancel{g\;adduct} \times \frac{1\;\cancel{mole\;adduct}}{162.28\;\cancel{g\;adduct}} \times \frac{2\;\cancel{moles\;bd}}{1\;\cancel{mole\;adduct}} \times \frac{54.09\;\cancel{g\;bd}}{1\;\cancel{mole\;bd}}$$

$$\times \frac{1000\;\cancel{g\;bd\;soln}}{150\;\cancel{g\;bd}} \times \frac{1\;mL\;bd\;soln}{0.682\;\cancel{g\;bd\;soln}} \approx 32.6\;mL\;bd\;soln$$

In the other calculation, the stoichiometry is 1:1. Moles of dimethylacetylene (abbreviated as "dma") may then be converted to mass, using its molar mass, and subsequently to volume, using its density.

$$5.00 \;\cancel{g\;adduct} \times \frac{1\;\cancel{mole\;adduct}}{162.28\;\cancel{g\;adduct}} \times \frac{1\;\cancel{moles\;dma}}{1\;\cancel{mole\;adduct}} \times \frac{54.09\;\cancel{g\;dma}}{1\;\cancel{mole\;dma}}$$

$$\times \frac{1\;mL\;dma}{0.691\;\cancel{g\;dma}} \approx 2.41\;mL\;dma$$

14. To calculate the percent yield, we must first determine the theoretical yield. This requires two calculations. The first begins with the volume of 1,3-butadiene solution (abbreviated as "bd soln") and uses its density to convert to mass of solution. Then, we use the conversion factor derived in Problem 13(b) (150 grams of 1,3-butadiene/1,000 grams of 1,3-butadiene solution) to convert mass of the solution to mass of 1,3-butadiene. The molar mass of 1,3-butadiene is used to obtain moles, and the mole ratio is 1:1. Finally, the products' (abbreviated as "p") molar mass enables us to convert moles to mass. In regard to the mole ratio used, remember the lesson of Problem 3: 1 mole of reactant can only become 1 mole of product here as well; the fact there is a choice of more than one constitutional isomer of the product is irrelevant.

$$6.75 \;\cancel{mL\;bd\;soln} \times \frac{0.682\;\cancel{g\;bd\;soln}}{1\;\cancel{mL\;bd\;soln}} \times \frac{150\;\cancel{g\;bd}}{1000\;\cancel{g\;bd\;soln}} \times \frac{1\;\cancel{mole\;bd}}{54.09\;\cancel{g\;bd}}$$

$$\times \frac{1\;\cancel{mole\;p}}{1\;\cancel{mole\;bd}} \times \frac{135.00\;g\;p}{1\;\cancel{mole\;p}} \approx 1.72\;g\;product$$

The second calculation begins with the volume of HBr solution (abbreviated as "HBr soln"). The steps mirror those used in the preceding calculation very closely, though we'll of course substitute the relevant values of density, molar mass, and the conversion factor derived in Problem 8(a).

$$2.25 \; mL \; HBr \; soln \times \frac{1.40 \; g \; HBr \; soln}{1 \; mL \; HBr \; soln} \times \frac{330 \; g \; HBr}{1000 \; g \; HBr \; soln} \times \frac{1 \; mole \; HBr}{80.91 \; g \; HBr}$$

$$\times \frac{1 \; mole \; p}{1 \; mole \; HBr} \times \frac{135.00 \; g \; p}{1 \; mole \; p} \approx 1.73 \; g \; product$$

The smaller of the two values, 1.72 grams, is the theoretical yield, and 1,3-butadiene was the limiting reactant. We can now ascertain that the percent yield, when 1.53 grams of the product mixture is obtained, is about 89.0%.

$$Percent \; Yield = \frac{amount \; of \; product \; obtained}{theoretical \; yield} \times 100$$

$$= \frac{1.53 \; g \; product}{1.72 \; g \; product} \times 100 \approx 89.0\%$$

15. First, we must make sure the equation is balanced. Remember that, as we've seen in Problems 3 and 14, there is no impact on the stoichiometry when more than one constitutional isomer of the product is possible. One molecule of toluene can become only one molecule of ethyltoluene, so 1 mole of toluene can become only 1 mole of ethyltoluene. Also, we've been told that the aluminum trichloride is a catalyst, so it is neither a reactant nor a product. Note that the HCl by-product has been omitted but has no impact on the calculation.

Planning this synthesis requires three calculations. In the first, the desired mass of ethyltoluene (abbreviated as "EtTol") is converted to moles using its molar mass. The mole ratio is 1:1. Then, the number of moles of toluene (abbreviated as "PhMe") is converted to a mass, using its molar mass, and to a volume, using its density.

$$5.00 \; g \; EtTol \times \frac{1 \; mole \; EtTol}{120.20 \; g \; EtTol} \times \frac{1 \; mole \; PhMe}{1 \; mole \; EtTol} \times \frac{92.14 \; g \; PhMe}{1 \; mole \; PhMe}$$

$$\times \frac{1 \; mL \; PhMe}{0.865 \; g \; PhMe} \approx 4.43 \; mL \; toluene$$

The second calculation looks the same through to the mole ratio. However, the ethyl chloride (abbreviated as "EtCl") is available as a solution, so once we have the desired number of moles, we divide by the molarity (moles solute/L solution) to obtain the needed volume of ethyl chloride solution (abbreviated as "EtCl soln"). Lastly, we convert liters to milliliters, a more convenient unit for this reaction.

$$5.00 \; \cancel{g \; EtTol} \times \frac{1 \; \cancel{mole \; EtTol}}{120.20 \; \cancel{g \; EtTol}} \times \frac{1 \; \cancel{mole \; EtCl}}{1 \; \cancel{mole \; EtTol}} \times \frac{1 \; \cancel{L} \; EtCl \; soln}{2.0 \; \cancel{mole \; EtCl}}$$
$$\times \frac{1000 \; mL}{1 \; \cancel{L}} \approx 20.8 \; mL \; EtCl \; soln$$

In the final calculation, we've been told to use a catalytic amount of aluminum trichloride, so we place the stated 0.1 molar equivalent into the mole ratio. Then, we convert moles of aluminum trichloride to grams using its molar mass.

$$5.00 \; \cancel{g \; EtTol} \times \frac{1 \; \cancel{mole \; EtTol}}{120.20 \; \cancel{g \; EtTol}} \times \frac{0.1 \; \cancel{mole \; AlCl_3}}{1 \; \cancel{mole \; EtTol}}$$
$$\times \frac{133.33 \; g \; AlCl_3}{1 \; \cancel{mole \; AlCl_3}} \approx 0.555 \; g \; AlCl_3$$

Before we finish, we must remember to adjust for the fact that this reaction has previously given us about 70% yield. Each amount of *reactant* will be divided by 0.70 to account for this. Note that the amount of catalyst need not necessarily be adjusted. Since catalysts are neither created nor destroyed in a chemical reaction, we could use the 0.555 grams of aluminum trichloride that our original planning called for. However, if you did scale up the amount of catalyst, that would not be wrong either.

$$\frac{4.43 \; mL \; toluene}{0.70} = 6.33 \; mL \; toluene$$

$$\frac{20.8 \; mL \; ethyl \; chloride \; solution}{0.70} = 29.7 \; mL \; ethyl \; chloride \; solution$$

16. The reaction equation is balanced as written. The notation "$-H_2O$" denotes that water, a by-product of the reaction, is being removed as it is formed. This is a freely reversible process, and the removal of water drives the reaction to completion. If it is more clear, you can also write the reaction equation with dimethylamine explicitly indicated as a reactant and water shown on the product side (Figure 10.6).

To calculate percent yield, we must first determine the theoretical yield. Cyclohexanone (abbreviated as "CHxone") is a neat liquid, so

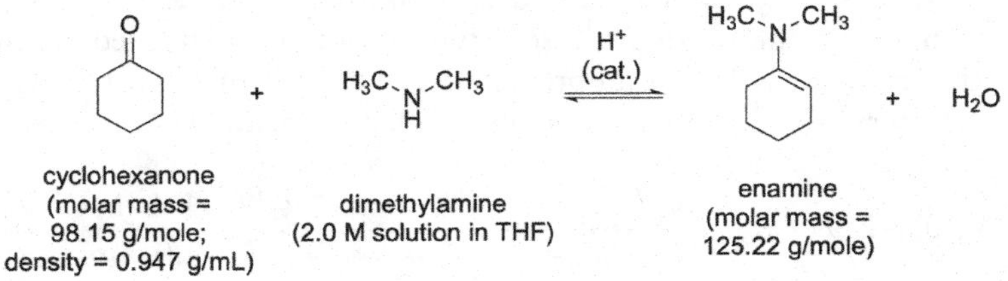

Figure 10.6 Alternate depiction of enamine formation from cyclohexanone.

the given volume is converted to moles using its density and molar mass. The mole ratio is 1:1, and the moles of enamine are converted to grams using its molar mass.

$$3.00 \; \cancel{mL \; CHxone} \times \frac{0.947 \; \cancel{g \; CHxone}}{1 \; \cancel{mL \; CHxone}} \times \frac{1 \; \cancel{mole \; CHxone}}{98.15 \; \cancel{g \; CHxone}}$$

$$\times \frac{1 \; \cancel{mole \; enamine}}{1 \; \cancel{mole \; CHxone}} \times \frac{125.22 \; g \; enamine}{1 \; \cancel{mole \; enamine}} \approx 3.62 \; g \; enamine$$

On the other hand, the dimethylamine (abbreviated as "dma") is provided as a solution (abbreviated as "soln"). Therefore, we first convert the volume in milliliters to liters. Then, the solution's molarity is used to obtain the number of moles of dimethylamine. Again, the mole ratio is merely 1:1. Finally, we obtain a mass of the enamine using its molar mass.

$$15.00 \; \cancel{mL \; dma \; soln} \times \frac{1 \; \cancel{L}}{1000 \; \cancel{mL}} \times \frac{2.0 \; \cancel{mole \; dma}}{1 \; \cancel{L \; dma \; soln}} \times \frac{1 \; \cancel{mole \; enamine}}{1 \; \cancel{mole \; dma}}$$

$$\times \frac{125.22 \; g \; enamine}{1 \; \cancel{mole \; enamine}} \approx 3.757 \; g \; enamine$$

The smaller of the two values, 3.62 grams, is the theoretical yield of enamine. If 3.51 grams of enamine are obtained, the percent yield is about 97.0%.

$$Percent \; Yield = \frac{amount \; of \; product \; obtained}{theoretical \; yield} \times 100$$

$$= \frac{3.51 \; \cancel{g \; enamine}}{3.62 \; \cancel{g \; enamine}} \times 100 \approx 97.0\%$$

17. We'll need to ascertain the theoretical yield first. The 2,2-dimethyl-propiophenone (abbreviated as "dmp") is a neat liquid, so we use its density and molar mass to obtain moles. The mole ratio is simply 1:1, and the molar mass of product (abbreviated as "tbb") is used to convert moles to a mass.

$$4.25 \text{ mL dmp} \times \frac{0.97 \text{ g dmp}}{1 \text{ mL dmp}} \times \frac{1 \text{ mole dmp}}{162.23 \text{ g dmp}} \times \frac{1 \text{ mole tbb}}{1 \text{ mole dmp}}$$
$$\times \frac{178.23 \text{ g tbb}}{1 \text{ mole tbb}} \approx 4.53 \text{ g tert} - \text{butyl benzoate}$$

For the mCPBA, we must first account for the fact that the sample contains only 77% of the desired reagent by multiplying by 0.77. Then, mCPBA's molar mass is used to convert to moles. The endgame for this calculation then mirrors what we did above.

$$6.00 \text{ g mCPBA} \times 0.77 \times \frac{1 \text{ mole mCPBA}}{172.57 \text{ g mCPBA}} \times \frac{1 \text{ mole tbb}}{1 \text{ mole mCPBA}}$$
$$\times \frac{178.23 \text{ g tbb}}{1 \text{ mole tbb}} \approx 4.77 \text{ g tert} - \text{butyl benzoate}$$

The smaller value, 4.53 grams, is the theoretical yield of *tert*-butyl benzoate, and 2,2-dimethylpropiophenone was the limiting reactant. When 4.50 grams of *tert*-butyl benzoate are obtained, the percent yield is an impressive 99.3%.

$$\text{Percent Yield} = \frac{\text{amount of product obtained}}{\text{theoretical yield}} \times 100$$
$$= \frac{4.50 \text{ g tbb}}{4.53 \text{ g tbb}} \times 100 \approx 99.3\%$$

18. Did you perform your calculations without first balancing the reaction equation? If so, return to the equation, balance it, and make the necessary adjustments to your arithmetic before proceeding.

Examination of the reaction equation shows two nitrogen atoms bearing methyl (CH_3) groups on the product side but only one in the reactants and reagents. Therefore, we must place a stoichiometric coefficient of 2 in front of methylamine to balance the equation (Figure 10.7).

Now, we can begin planning the synthesis. Two calculations will be required. In the first, we begin with the desired mass of N-methylbenzamide (abbreviated as "nmb"). Its molar mass allows

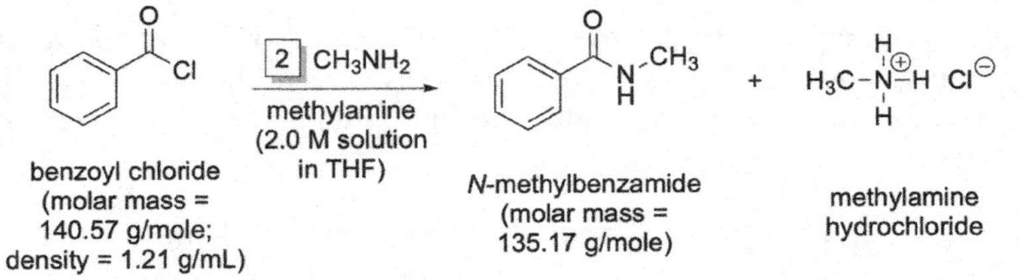

Figure 10.7 Balanced amide formation from benzoyl chloride.

conversion to moles, and the mole ratio is simply 1:1. The needed moles of benzoyl chloride (abbreviated as "bc") are converted to a volume using its molar mass and density because it is a neat liquid.

$$20.00 \; g \; nmb \times \frac{1 \; mole \; nmb}{135.17 \; g \; nmb} \times \frac{1 \; mole \; bc}{1 \; mole \; nmb} \times \frac{140.57 \; g \; bc}{1 \; mole \; bc}$$

$$\times \frac{1 \; mL \; bc}{1.21 \; g \; bc} \approx 17.19 \; mL \; benzoyl \; chloride$$

On the other hand, methylamine (abbreviated as "ma") is provided as a solution (abbreviated as "soln"). Once we've obtained moles of N-methylbenzamide, the mole ratio is 1 mole N-methylbenzamide to 2 moles methylamine. Then, the needed moles of methylamine are converted to a volume of methylamine solution using the solution's concentration in molarity (moles solute/L solution). Lastly, the volume in liters is converted to milliliters for the sake of convenience.

$$20.00 \; g \; nmb \times \frac{1 \; mole \; nmb}{135.17 \; g \; nmb} \times \frac{2 \; mole \; ma}{1 \; mole \; nmb} \times \frac{1 \; L \; ma \; soln}{2.0 \; mole \; ma}$$

$$\times \frac{1000 \, mL}{1 \, L} \approx 148.0 \, mL \; methylamine \; solution$$

19. Did you remember to balance the reaction before beginning? If not, do that and adjust your calculations accordingly before proceeding.

    Inspection of the reaction equation reveals that two molar equivalents of methylmagnesium bromide will be needed to balance the reaction because two methyl (CH_3) groups have been added to the product (Figure 10.8).

    With the balanced equation in hand, planning can begin. The first calculation starts with the desired amount of 2-phenyl-2-propanol (abbreviated as "PhPrOH"). Its molar mass is used to effect conversion to moles, and the mole ratio for this computation is 1:1. The

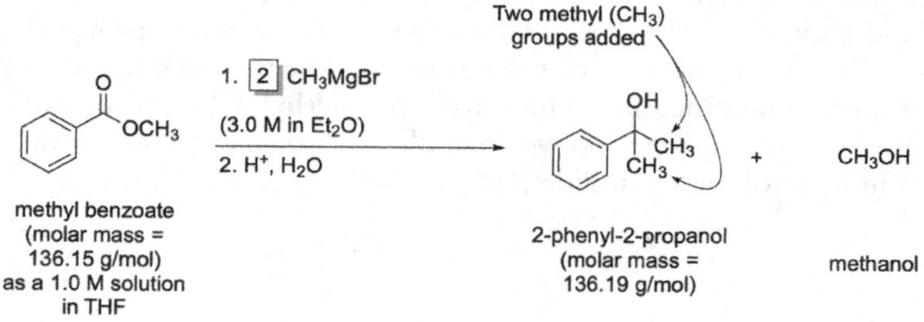

Figure 10.8 Balanced Grignard reaction of methyl benzoate.

needed moles of methyl benzoate (abbreviated as "mb") are converted to a volume of the methyl benzoate solution (abbreviated as "mb soln") using the solution's molarity (moles solute/L solution). Lastly, the volume is converted from liters to milliliters for convenience.

$$5.50 \; \cancel{g \; PhPrOH} \times \frac{1 \; \cancel{mole \; PhPrOH}}{136.19 \; \cancel{g \; PhPrOH}} \times \frac{1 \; \cancel{mole \; mb}}{1 \; \cancel{mole \; PhPrOH}} \times \frac{1 \; \cancel{L} \; mb \; soln}{1.0 \; \cancel{mole \; mb}}$$

$$\times \frac{1000 \; mL}{1 \; \cancel{L}} \approx 40.4 \; mL \; methyl \; benzoate \; solution$$

The second calculation begins in the same fashion, but this time, the mole ratio is 1 mole 2-phenyl-2-propanol to 2 moles methylmagnesium bromide (abbreviated as "MeMgBr"). The solution's molarity is used to convert the necessary number of moles to a volume of solution (abbreviated as "soln"), and once again, the volume is converted to milliliters.

$$5.50 \; \cancel{g \; PhPrOH} \times \frac{1 \; \cancel{mole \; PhPrOH}}{136.19 \; \cancel{g \; PhPrOH}} \times \frac{2 \; \cancel{mole \; MeMgBr}}{1 \; \cancel{mole \; PhPrOH}}$$

$$\times \frac{1 \; \cancel{L} \; MeMgBr \; soln}{3.0 \; \cancel{mole \; MeMgBr}} \times \frac{1000 \; mL}{1 \; \cancel{L}}$$

$$\approx 26.9 \; mL \; MeMgBr \; solution$$

Finally, if we anticipate a 65% yield, then the amount of each reactant must be scaled up by dividing by 0.65.

$$\frac{40.4 \; mL \; methyl \; benzoate \; solution}{0.65} = 62.2 \; mL \; methyl \; benzoate \; solution$$

$$\frac{26.9 \; mL \; MeMgBr \; solution}{0.65} = 41.4 \; mL \; MeMgBr \; solution$$

20. If you neglected to balance the equation before beginning, go back and make the needed corrections to your work before proceeding.

First, we must balance the equations. The first step is the only one requiring modification. There are three methyl (CH_3) groups added to the reactant in step 1, so we must place a stoichiometric coefficient of 3 in front of methyl iodide (Figure 10.9).

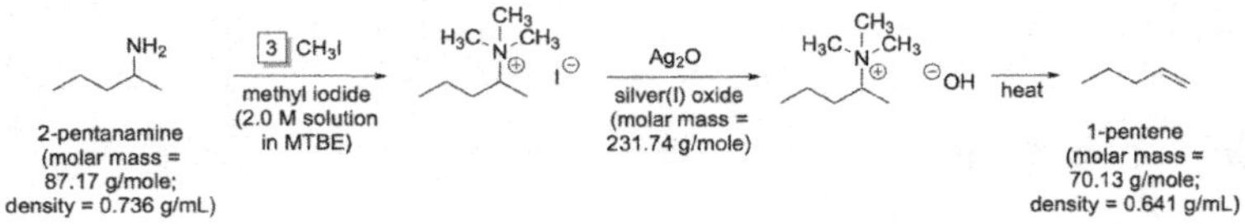

Figure 10.9 Balanced Hofmann elimination of 2-pentanamine.

Now, we can begin planning. We'll need to calculate the necessary amounts of 2-pentanamine, methyl iodide, and silver(I) oxide. We do not need to calculate amounts of any of the synthetic intermediates because they are produced from the original reactant, 2-pentanamine.

The first calculation uses the density and molar mass of 1-pentene to convert the desired volume to moles. The mole ratio of 1-pentene to 2-pentanamine (abbreviated as "pa") is 1:1. Then, the molar mass and density of 2-pentanamine are used to convert the needed number of moles to a volume.

$$7.50 \text{ mL pentene} \times \frac{0.641 \text{ g pentene}}{1 \text{ mL pentene}} \times \frac{1 \text{ mole pentene}}{70.13 \text{ g pentene}} \times \frac{1 \text{ mole pa}}{1 \text{ mole pentene}}$$

$$\times \frac{87.17 \text{ g pa}}{1 \text{ mole pa}} \times \frac{1 \text{ mL pa}}{0.736 \text{ g pa}} \approx 8.12 \text{ mL pa}$$

The second calculation is the same through to the mole ratio, which differs in this case. It is one mole 1-pentene to three moles methyl iodide (MeI). Then, the molarity of the methyl iodide solution (abbreviated as "MeI soln") is used to convert the necessary number of moles to a volume of the solution. Finally, conversion of volume in liters to milliliters is convenient.

$$7.50 \text{ mL pentene} \times \frac{0.641 \text{ g pentene}}{1 \text{ mL pentene}} \times \frac{1 \text{ mole pentene}}{70.13 \text{ g pentene}} \times \frac{3 \text{ mole MeI}}{1 \text{ mole pentene}}$$

$$\times \frac{1 \text{ L MeI soln}}{2.0 \text{ mole MeI}} \times \frac{1000 \text{ mL}}{1 \text{ L}} \approx 103 \text{ mL MeI soln}$$

In the last calculation, the number of moles of 1-pentene is calculated identically, but here the mole ratio is 1:1. Then, the needed moles of silver(I) oxide are converted to a mass using its molar mass.

$$7.50 \; \cancel{mL \; pentene} \times \frac{0.641 \; \cancel{g \; pentene}}{1 \; \cancel{mL \; pentene}} \times \frac{1 \; \cancel{mole \; pentene}}{70.13 \; \cancel{g \; pentene}}$$

$$\times \frac{1 \; \cancel{mole \; Ag_2O}}{1 \; \cancel{mole \; pentene}} \times \frac{231.74 \; g \; Ag_2O}{1 \; \cancel{mole \; Ag_2O}} \approx 15.9 \; g \; Ag_2O$$

Our final step is to account for the expected 90% overall yield by scaling up accordingly.

$$\frac{8.12 \; mL \; 2-pentanamine}{0.90} = 9.02 \; mL \; 2-pentanamine$$

$$\frac{103 \; mL \; MeI \; soln}{0.90} = 114 \; mL \; MeI \; soln$$

$$\frac{15.9 \; g \; Ag_2O}{0.90} = 17.7 \; g \; Ag_2O$$

Index

Printed in the United States
by Baker & Taylor Publisher Services